CONVECTIVE INSTABILITIES
IN SYSTEMS WITH INTERFACE

CONVECTIVE INSTABILITIES
IN SYSTEMS WITH INTERFACE

Ilya B. Simanovskii
Perm Pedagogical Institute, Russia

and

Alexandre A. Nepomnyashchy
Technion Israel Institute of Technology, Haifa

GORDON AND BREACH SCIENCE PUBLISHERS
Switzerland • Australia • Belgium • France • Germany • Great Britain
India • Japan • Malaysia • Netherlands • Russia • Singapore • USA

Gordon and Breach Science Publishers

Y-Parc
Chemin de la Sallaz
CH-1400 Yverdon, Switzerland

Post Office Box 90
Reading, Berkshire RG1 8JL
Great Britain

Private Bag 8
Camberwell, Victoria 3124
Australia

3-14-9, Okubo
Shinjuku-ku, Tokyo 169
Japan

58, rue Lhomond
75005 Paris
France

Emmaplein 5
1075 AW Amsterdam
Netherlands

Glinkastrasse 13-15
O-1086 Berlin
Germany

820 Town Center Drive
Langhorne, Pennsylvania 19047
United States of America

Library of Congress Cataloging-in-Publication Data

Simanovskii, I. B., 1951 –
    Convective instabilities in systems with interface / I. B.
Simanovskii, A. A. Nepomnyashchy.
        p.   cm.
    Includes bibliographical references and index.
    ISBN 2-88124-924-8
    1. Heat--Convection. I. Nepomnyashchy, A. A., 1950-   .
    II. Title.
TJ260.S568   1993
621.402'2--dc20

93-43821
CIP

Contents

PREFACE

Convective phenomena taking place in a system with an interface attract much interest. First of all, these phenomena greatly influence the heat-mass transfer intensity used in different technological processes. It is quite sufficient to recall such examples as: mass transfer problems in chemical technology [149], production of superpure melts under reduced gravity conditions [7, 127], liquid hydrocarbon underground storage [83, 170]. These are practical applications of surface convection theory. In addition, geophysical applications of this theory are also important [3, 4, 25, 43, 64, 177].

Surface convection is characterized by a large variety of physical effects, which essentially influence hydrodynamic stability and heat-mass transfer processes.

Besides the well-known case of classical Rayleigh instability mechanism, thermocapillary effects play an essential role in interface destabilization. When the media are solutions, the convective mechanisms connected with density and surface tension depend on the concentration ingredients. Convective generation is greatly influenced by adsorbed surfactants, heat release or heat absorption on the interface and surface deformation. A specific instability mechanism is realized for some systems when heating from above [49, 187]. The variety of instability mechanisms makes convection in systems with an interface an interesting case for new ideas and approaches of modern hydrodynamic stability theory. Up to now mainly original papers contained the results of surface convection investigations. Problems connected with concentration-capillary and concentration-gravity convection are presented to a certain extent in collected papers [78]. In this book we try to systematize the results on heat convection in the presence of an interface. Attention will mainly focus on the results presented in papers concerning the problem in a fully consistent treatment, where heat and hydrodynamic interaction on the interface are taken into account.

A unified notation system is used which does not always coincide with those used in previously published papers.

CHAPTER 1

Introduction

1.1. A consistent approach to surface phenomena investigations

Starting with the first papers devoted to surface convection, two main approaches have been used. The first one aimed at effects on the gas-fluid interface, was based on the neglect of hydrodynamic and heat processes in gas media. Only fluid phase effects were investigated; the interface has been assumed to be free and empiric coefficients like the Biot number have been used to describe interface temperature boundary conditions of the third kind. Relative mathematical simplicity of this approach gave an opportunity of solving some fundamental problems basing on it. Some of these problems are: thermocapillary instability of a layer with free interface investigation [119, 124], analysis of the effects connected with an interface deformation [46, 157], investigation of the surfactant effect on thermocapillary and thermogravitational instability [13, 123].

Today a monofluid approach is the main one whenever convective phenomena numerical models in systems with a gas-fluid interface are meant[1] [127, 168].

Another approach was suggested for analysing convection in immiscible fluids systems. In this last, convection equations are laid down for both the media together with boundary conditions on the interface.

On this basis a new instability effect in a two-layer system was discovered in an early paper of Sternling and Scriven [174]. Then a series of new effects essentially connected with hydrodynamic and heat interaction on the media interface has been discovered. To analyse them, another adjoint approach is to be used: "fracture" character of monotonous neutral curves at thermocapillary and concentration-capillary instability [169], specific non-Rayleigh mechanism of thermogravitational instability when

---

[1] Let us also note the simplified approach used in papers [120, 121]. Heat and mass exchange being calculated, full nonlinear equations in adjoint statement are used and velocity fields are approximated on the basis of linearized Navier-Stokes equations system solution. But convective motion intensity can not be found within the given method frames. So it is to be estimated using additional considerations (for example, the energy balance condition).

1

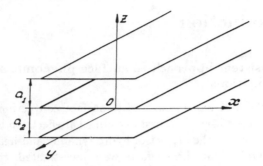

Figure 1.1. A system of horizontal layers.

heating from above [49,187], oscillatory thermogravitational instability when heating from below [50].

It is worth noting that even for a gas-fluid system the monofluid approach is not closed and consistent, because it requires an empiric coefficient, the Biot number which is assumed to be constant on the whole free surface.

The given work presents investigation results, convection being studied on the basis of adjoint approach.

1.2.   Mathematical statement of the problem

1. Equations and boundary conditions. Properties of surface convection can be considered if a system of two flat horizontal layers of immiscible fluids with different physical properties is taken as an example (see Fig. 1.1). The system is limited from above and from below by two isothermic plates kept at constant nonequal temperatures. Let us choose the upper boundary temperature for the reference one; in this case the lower boundary temperature ϑ can be both positive (heating from below) and negative (heating from above). Densities, dynamic and kinematic viscosities, heat conductivity, temperature diffusivity and volume expansion coefficients are ρ_m, η_m, ν_m, κ_m, χ_m, β_m, $m = 1, 2$. Henceforwards all the values referring to the upper fluid are marked with index 1, and the values referring to the lower fluid are marked with index 2.

Let us assume that the surface tension on the interface depends on temperature : $\sigma = \sigma(T)$. If temperature drops are not too large then this dependence can be considered linear : $\sigma = \sigma_0 - \alpha T$ (σ_0 and α are constants). The case when at some temperature the surface tension shows a minimum (anomalous thermocapillary effect) is an exception. The case has been observed for some solutions. Under such conditions, convection has certain specific peculiarities [134, 95, 183, 184]. Investigations of this problem are far from being completed and we are not going to deal with them in this book, referring the reader to the review [135].

Let us introduce a coordinate system as shown in Fig. 1.1, directing axes x and y horizontally and axis z—vertically upward. Solid boundaries are set by equations $z = a_1$ and $z = -a_2$. The interface between both fluids is described by equation

$$z = h(x, y, t), \qquad (1.1)$$

where function h is to be defined within the problem solution process.

Let us write down equations of convection in both fluids using Boussinesq approximation [48]:

$$\frac{\partial v_m}{\partial t} + (v_m \nabla) v_m = -\frac{1}{\rho_m} \nabla p_m + \nu_m \Delta v_m + g \beta_m T_m \gamma,$$

$$\frac{\partial T_m}{\partial t} + v_m \nabla T_m = \chi_m \Delta T_m, \quad \operatorname{div} v_m = 0; \quad m = 1, 2. \qquad (1.2)$$

Here g is gravity, γ is a unit vector, directed vertically upward. Let us remind of the fact that in equations (1.2) p_m is the hydrostatic pressure; the total pressure equals to $p_m - \rho_m gz$. The conditions on the solid isothermic walls are :

$$z = a_1 : \quad v_1 = 0, \quad T_1 = 0; \qquad (1.3)$$
$$z = -a_2 : \quad v_2 = 0, \quad T_2 = \vartheta. \qquad (1.4)$$

To write down boundary conditions on the interface (1.1), we introduce normal vector n and orthogonal tangential vectors $\tau^{(1)}$, $\tau^{(2)}$. The balance condition of normal stresses is presented in [92]:

$$z = h : \quad (p_1 - \rho_1 gh) - (p_2 - \rho_2 gh) + \frac{\sigma}{R} = (\sigma_{1,ik} - \sigma_{2,ik}) n_i n_k. \qquad (1.5)$$

For tangential stresses we get two conditions:

$$z = h : \quad (\sigma_{1,ik} - \sigma_{2,ik}) \tau_i^{(l)} n_k - \alpha \tau_i^{(l)} \frac{\partial T_1}{\partial x_i} = 0, \quad l = 1, 2. \qquad (1.6)$$

Here R is the surface curvature radius, $\sigma_{m,ik} = \eta_m(\partial v_{m,i}/\partial x_k + \partial v_{m,k}/\partial x_i)$ is the viscous stress tensor of the m-th fluid.

Also, continuity of the velocity field takes place along the interface:

$$z = h: \quad v_1 = v_2. \tag{1.7}$$

Interface shape transformation in time is described by the kinematic condition

$$z = h: \quad \frac{\partial h}{\partial t} + v_{1,x}\frac{\partial h}{\partial x} + v_{1,y}\frac{\partial h}{\partial y} = v_{1,z}. \tag{1.8}$$

Heat conditions on the interface are: temperature fields continuity

$$z = h: \quad T_1 = T_2 \tag{1.9}$$

and heat flow normal components continuity

$$z = h: \quad \left(\kappa_1\frac{\partial T_1}{\partial x_i} - \kappa_2\frac{\partial T_2}{\partial x_i}\right)n_i = 0. \tag{1.10}$$

Let us introduce the following notation: $\eta = \eta_1/\eta_2$, $\nu = \nu_1/\nu_2$, $\kappa = \kappa_1/\kappa_2$, $\chi = \chi_1/\chi_2$, $\beta = \beta_1/\beta_2$, $a = a_2/a_1$. We shall go over to dimensionless variables choosing for length, time, velocity, pressure and temperature units respectively $a_1, a_1^2/\nu_1, \nu_1/a_1, \rho_1\nu_1^2/a_1^2$ and $|\vartheta|$. The previous equations and boundary conditions take the form:

$$\frac{\partial v_1}{\partial t} + (v_1\nabla)v_1 = -\nabla p_1 + \Delta v_1 + GT_1\gamma, \tag{1.11}$$

$$\frac{\partial T_1}{\partial t} + v_1\nabla T_1 = \frac{1}{P}\Delta T_1, \quad \mathrm{div}\,v_1 = 0$$

$$\frac{\partial v_2}{\partial t} + (v_2\nabla)v_2 = -\rho\nabla p_2 + \frac{1}{\nu}\Delta v_2 + \frac{G}{\beta}T_2\gamma, \tag{1.12}$$

$$\frac{\partial T_2}{\partial t} + v_2\nabla T_2 = \frac{1}{\chi P}\Delta T_2, \quad \mathrm{div}\,v_2 = 0;$$

$$z = 1: \quad v_1 = 0, \quad T_1 = 0, \tag{1.13}$$

$$z = -a: \quad v_2 = 0, \quad T_2 = s; \tag{1.14}$$

$$z = h: p_1 - p_2 + \frac{W_0}{R}(1 - \delta_\alpha T) + Ga(\rho^{-1} - 1)h =$$
$$\left[\left(\frac{\partial v_{1,i}}{\partial x_k} + \frac{\partial v_{1,k}}{\partial x_i}\right) - \eta^{-1}\left(\frac{\partial v_{2,i}}{\partial x_k} + \frac{\partial v_{2,k}}{\partial x_i}\right)\right] n_i n_k, \tag{1.15}$$

$$\left[\eta\left(\frac{\partial v_{1,i}}{\partial x_k} + \frac{\partial v_{1,k}}{\partial x_i}\right) - \left(\frac{\partial v_{2,i}}{\partial x_k} + \frac{\partial v_{2,k}}{\partial x_i}\right)\right] \tau_i^{(l)} n_k -$$
$$- Mr\tau_i^{(l)}\frac{\partial T_1}{\partial x_i} = 0, \; l = 1, 2, \tag{1.16}$$

$$v_1 = v_2, \tag{1.17}$$

$$\frac{\partial h}{\partial t} + v_{1,x}\frac{\partial h}{\partial x} + v_{1,y}\frac{\partial h}{\partial y} = v_{1,z}, \tag{1.18}$$

$$T_1 = T_2, \tag{1.19}$$

$$\left(\kappa\frac{\partial T_1}{\partial x_i} - \frac{\partial T_2}{\partial x_i}\right) n_i = 0. \tag{1.20}$$

Here $G = g\beta_1|\vartheta|a_1^3/\nu_1^2$ is the Grashof number, $P = \nu_1/\chi_1$ is the Prandtl number, $Mr = \alpha|\vartheta|a_1/\eta_2\nu_1$ is the Marangoni number analogue, $Ga = ga_1^3/\nu_1^2$ is the Galileo number, $W_0 = \sigma_0 a_1/\eta_1\nu_1$, $\delta_\alpha = \alpha|\vartheta|/\sigma_0$. The value of $s = \text{sign}\vartheta$ determines the heating character: $s = 1$ for heating from below and $s = -1$ for heating from above.

2. Influence of convective flow on an interface shape. A peculiarity of the formulated boundary problem is the fact that the interface shape is not known beforehand and is to be found simultaneously with velocity, pressure and temperature fields. There are nevertheless some important cases when convective flow influence on the interface shape can be neglected.

First of all let us note that equations of convection in the Boussinesq approach are derived with the assumption that the parameter $\delta_\alpha = \beta_1|\vartheta| = G/Ga$ is supposed to be small [48]. That is why these equations can be used for thermogravitational convection ($G \neq 0$) only in case such that $Ga \gg G$. Within the limit $Ga \to \infty$ the terms connected with convective motion in the boundary condition (1.15) must be omitted. In this case the interface shape is determined by the balance of normal stresses having hydrostatic and capillary origin [92]:

$$(\rho^{-1} - 1)h + \frac{r_c^2}{R}(1 - \delta_\alpha T) = C \qquad (1.21)$$

where $r_c = (W_0/Ga)^{1/2} = (\sigma_0/\rho_1 g a_1^2)^{1/2}$ is the dimensionless capillary radius, constant C characterizes stresses difference on both the sides of the interface. Media with close densities ($\rho \simeq 1$) at small capillary radius ($r_c \ll 1$) are an exception and in this case the problem is to be considered in full statement.

For infinite layers system $R = \infty$ so that it follows from equation (1.21) that

$$h = 0 \qquad (1.22)$$

(1.18) being written as

$$v_{1,z} = 0. \qquad (1.23)$$

In case the motion takes place in a closed cavity, the interface shape $z = h(x, y)$ is determined from (1.21), contact angles close to solid walls being taken into account. If the surface tension change $\alpha|\vartheta|$, connected with temperature dependence, is small in comparison with its mean value σ_0 (as it is the case when linear approximation for $\sigma(T)$ is used) then the term including δ_α may be neglected in formula (1.21). In this case the interface determined by

$$(\rho^{-1} - 1)h + \frac{r_c^2}{R} = C \qquad (1.24)$$

is stable and equality (1.23) is fulfilled.

The interface shape described by equation (1.24) is the zeroth approximation for the full problem solution with respect to the parameters δ_β, δ_α. Velocity, pressure and temperature fields found on the basis of boundary problem with the stable interface solution can be used to calculate the next approximation including corrections proportional to small parameters.

For example, for a system of infinite layers, from the boundary condition (1.15) follows an equation

$$(\rho^{-1} - 1)h + r_c^2 \left(\frac{\partial^2 h}{\partial x^2} + \frac{\partial^2 h}{\partial y^2} \right) =$$

$$= \delta_\beta G^{-1} \left[p_2 - p_1 + 2 \left(\frac{\partial v_{1,z}}{\partial z} - \eta^{-1} \frac{\partial v_{2,z}}{\partial z} \right) \right] \Bigg|_{z=0} \quad (1.25)$$

determining surface relief induced by convective motion. Let us stress that there must not be any calculation of the interface distortion reverse influence on convective flow in system (1.11)–(1.20) because it is of the order of dominating as the summands omitted when the equations were derived within the Boussinesq approximation.

Let us now discuss the case of Galileo number finite values. In the framework of the Boussinesq approximation ($\delta_\beta \to 0$) and thus $G \to 0$, so only thermocapillary convection may take place. This situation can be realized for thin layers or under reduced gravity conditions.

If the value δ_α is small and Mr number is finite, then the most important summand of equation (1.15) is the term containing the parameter $W_0 = Mr\delta_\alpha^{-1}\eta^{-1}$. In this case the interface is not essentially influenced by thermocapillary motion. As in the case of thermogravitational convection, the relief is determined by hydrostatic and capillary forces balance and is described by equation (1.24) from which equality (1.23) follows.

The longwave disturbances of the interface for which the summand W_0/R is not a dominating one due to smallness of surface curvature are a special case. In this case the problem is to be considered in full statement. So, the dependence of the interface shape on convective motion must be taken into account in the following cases:

1) for thermogravitational convection in the case where use of the Boussinesq approximation ($\delta_\beta \geq 1$) is not valid;

2) for thermocapillary convection in a system of two media with close densities ($|\rho - 1| \ll 1$) at a small capillary radius ($r_c \ll 1$);

3) for both types of convection at strong dependence of the surface tension on temperature ($\delta_\alpha \geq 1$);

4) for a longwave thermocapillary convection.

In other cases both thermogravitational and thermocapillary convection can be considered without taking into consideration the convective motion influence on the interface shape.

3. Curvilinear coordinates. If the form of the vessel a container in which fluid movement takes place differs from a rectangular one, or if the interface is curved, a special system of curvilinear coordinates ξ^1, ξ^2 is natural. This coordinate system is constructed in such a way that an interface is described by the equation $\xi^2 = 0$ (the interface is supposed to be steady). In curvilinear coordinates convection equations (1.11), (1.12) get the form [131]

$$\frac{\partial v_1^i}{\partial t} + \frac{1}{\sqrt{g}}(\sqrt{g}v_1^i v_1^j)_{,j} = -g^{ij}p_{1,j} +$$

$$\frac{1}{\sqrt{g}}(\sqrt{g}d_1^{ij})_{,j} + GT_1\gamma^i, \tag{1.26}$$

$$\frac{\partial T_1}{\partial t} + \frac{1}{\sqrt{g}}(\sqrt{g}v_1^i T_1)_{,i} = \frac{1}{P\sqrt{g}}(\sqrt{g}g^{ij}T_{1,j})_{,i} \quad (\sqrt{g}v_1^i)_{,i} = 0,$$

$$\frac{\partial v_2^i}{\partial t} + \frac{1}{\sqrt{g}}(\sqrt{g}v_2^i v_2^j)_{,j} = -\rho g^{ij}p_{2,j} + \frac{1}{\nu\sqrt{g}}(\sqrt{g}d_2^{ij})_{,j} + \frac{G}{\beta}T_2\gamma^i,$$

$$\frac{\partial T_2}{\partial t} + \frac{1}{\sqrt{g}}(\sqrt{g}v_2^i T_2)_{,i} = \frac{1}{\chi P\sqrt{g}}(\sqrt{g}g^{ij}T_{2,j})_{,i} \quad (\sqrt{g}v_2^i)_{,i} = 0.$$

Here $g_{ij} = \frac{\partial x}{\partial \xi_i}\frac{\partial x}{\xi_j} + \frac{\partial z}{\partial \xi_j}\frac{\partial z}{\partial \xi_j}, g^{ij}$ are co- and contravariant components of the metric tensor, g — being the metric tensor determinant,

$$d_m^{ij} = v_{m;k}^j g^{ki} + v_{m;k}^i g^{kj} \quad (m = 1, 2).$$

The notation $,j$ stands for partial derivative and the notation $;j$ stands for covariant one. Solid boundaries conditions keep the form (1.13), (1.14).

Let us write the coditions at the interface $\xi^2 = 0$:

$$p_1 - p_2 + \frac{W_0}{R}(1 - \delta_\alpha T) + Ga(\rho^{-1} - 1)h =$$

$$\frac{g}{\sqrt{g}_{11}}d_1^{22} - \eta^{-1}\frac{g}{\sqrt{g}_{11}}d_2^{22}, \tag{1.27}$$

$$\eta\sqrt{g}(d_1^{21} - \frac{g_{12}}{g_{11}}d_1^{22}) - \sqrt{g}(d_2^{21} - \frac{g_{12}}{g_{11}}d_2^{22}) - MrT_{1,1}(\sqrt{g}_{11})^{-1} = 0,$$

$$v_1^1 = v_2^1, \quad v_1^2 = v_2^2 = 0, \quad T_1 = T_2,$$

$$\frac{\kappa}{\sqrt{g}}\left(T_{1,2} - \frac{g_{12}}{g_{11}}\right) = \frac{1}{\sqrt{g}}\left(T_{2,2} - \frac{g_{12}}{g_{11}}T_{2,1}\right).$$

Using these boundary conditions, one must keep in mind that the metric tensor may show a discontinuity on the interface if such a curvilinear coordinate system is considered.

4. The problem modifications. Described in this paragraph system of equations and boundary conditions (1.2)–(1.10) are traditionally applied to investigate convection in case there is an interface. But it is worth mentioning that some system modifications are considered in details in references.

In papers [135, 98] a more precise definition of boundary condition (1.10) is presented, energetic expenditures on an interface deformation being taken into consideration:

$$\kappa_2 \frac{\partial T_2}{\partial n} - \kappa_1 \frac{\partial T_1}{\partial n} = \alpha T_1 \text{div}_{\text{II}} v_1 + w \left(\frac{\partial T_1}{\partial t} + v_1 \nabla T_1 \right),$$

where $\partial/\partial n = n\nabla, \text{div}_{\text{II}} = \nabla - (n\nabla)n, \alpha = -d\sigma/dT$; the parameter $w = \frac{d}{dT}(\sigma + \alpha T)$ is not equal to zero if the function $\sigma(T)$ differs from a linear one. As shown in [135] the right–hand side ratio introduces a dimensionless parameter $\Lambda = \alpha^2 T_0/\eta_2 \chi_2$ which is small under realistic conditions (T_0 is the absolute interface temperature). Later on corrections to boundary condition (1.10) are not taken into consideration.

Some papers suggest taking into account additional dissipation mechanisms connected with effects on the interface ("surface viscosity"). For example, in paper [66] a general theory of surface viscosity is presented. Four phenomenological coefficients $\kappa, \eta, \kappa_N, \eta_N$ are used in it. In the majority of cases [78, 156] only two coefficients are used: surface displacement viscosity and expansion viscosity. Experimental papers devoted to surface viscosity measurement are not numerous up to now (see for example [37]). There is also a point of view according to which it is not necessary to introduce additional coefficients in boundary conditions, since the surface viscosity effects take place in reality, due to surfactants on the interface [94]. In the following sections, surface viscosity effects will not be taken into consideration.

CHAPTER 2

Convective Stability of a Two-Layer System

2.1. General properties of the problem for normal perturbations

1. Statement of the problem. In the previous chapter we began to consider the case of convection in a system of two flat horizontal layers containing immiscible fluids with different physical characteristics. The problem, whose formulation is summarized by (1.11)–(1.20), describes the fluid movement induced by the action of thermogravity and thermocapillarity.

Problem (1.11)–(1.20) for any choice of the parameters has the solution

$$
v_1^0 = v_2^0 = 0, \quad h^0 = 0,
$$
$$
T_1^0 = -\frac{s(z-1)}{1+\kappa a}, \quad T_2^0 = -\frac{s(\kappa z - 1)}{1 + \kappa a},
$$
$$
p_1^0 = -\frac{sG}{1+\kappa a}\left(\frac{z^2}{2} - z\right), \tag{2.1}
$$
$$
p_2^0 = -\frac{sG}{\rho\beta(1+\kappa a)}\left(\frac{\kappa z^2}{2} - z\right)
$$

corresponding to mechanical equilibrium.

This chapter is devoted to the study of mechanical stability a two-layer system within the limits of linear theory. Let us impose an infinitesimal disturbance on the reference state (2.1)

$$
(v_1, v_2, p_1, p_2, T_1, T_2, h).
$$

The boundary problem describing the effects of disturbance can be expressed through linearization of the full nonlinear problem (1.11)–(1.20) close to solution (2.1).

Dimensionless equations characterizing the effects of small disturbances can be written in the form:

$$\frac{\partial v_m}{\partial t} = -e_m \nabla p_m + c_m \Delta v_m + b_m GT_m \gamma, \tag{2.2}$$

$$\frac{\partial T_m}{\partial t} + A_m(v_m \gamma) = \frac{d_m}{P} \Delta T_m, \tag{2.3}$$

$$\mathrm{div} v_m = 0, \quad (m = 1, 2), \tag{2.4}$$

where $c_1 = b_1 = d_1 = e_1 = 1$, $c_2 = 1/\nu$, $b_2 = 1/\beta$, $d_2 = 1/\chi$, $e_2 = \rho$; $A_1 = dT_1^0/dz = -s/(1 + \kappa a)$, $A_2 = dT_2^0/dz = -s\kappa/(1 + \kappa a)$ are the dimensionless temperature gradients.

Conditions on the upper and the lower system boundaries are:

$$z = 1: \quad v_1 = 0, \quad T_1 = 0, \tag{2.5}$$
$$z = -a: \quad v_2 = 0, \quad T_2 = 0. \tag{2.6}$$

The linearized conditions (1.15)–(1.20) valid along the surface $z = h$ give rise to the following set of boundary conditions:

$$z = 0: p_1 - p_2 + Ga\left[(\rho^{-1} - 1) + \delta_\beta \frac{s(1 - \rho^{-1}\beta^{-1})}{1 + \kappa a}\right] h -$$
$$- W\Delta_2 h = 2\left(\frac{\partial v_{1,z}}{\partial z} - \eta^{-1}\frac{\partial v_{2,z}}{\partial z}\right), \tag{2.7}$$

$$\eta\left(\frac{\partial v_{1,x}}{\partial z} + \frac{\partial v_{1,z}}{\partial x}\right) - \left(\frac{\partial v_{2,x}}{\partial z} + \frac{\partial v_{2,z}}{\partial x}\right) - Mr\left(\frac{\partial T_1}{\partial x} + \frac{dT_1^0}{dz}\frac{\partial h}{\partial x}\right) = 0,$$
$$\eta\left(\frac{\partial v_{1,y}}{\partial z} + \frac{\partial v_{1,z}}{\partial y}\right) - \left(\frac{\partial v_{2,y}}{\partial z} + \frac{\partial v_{2,z}}{\partial y}\right) - Mr\left(\frac{\partial T_1}{\partial y} + \frac{dT_1^0}{dz}\frac{\partial h}{\partial y}\right) = 0,$$
$$\tag{2.8}$$

$$v_1 = v_2, \tag{2.9}$$
$$\frac{\partial h}{\partial t} = v_{1,z}, \tag{2.10}$$
$$T_1 - T_2 = \frac{s(1 - \kappa)}{1 + \kappa a} h, \tag{2.11}$$
$$\kappa \frac{\partial T_1}{\partial z} - \frac{\partial T_2}{\partial z} = 0. \tag{2.12}$$

where $\Delta_2 = \partial^2/\partial x^2 + \partial^2/\partial y^2$, $W = W_0(1 - \delta_\alpha s/(1 + \kappa a))$.

Since coefficients in the equations (2.2)–(2.4) and boundary conditions (2.5)–(2.12) do not depend on time and coordinates, the boundary problem solutions can be presented as a superposition of so-called "normal" disturbances characterized by a wave vector $k = (k_x, k_y)$ and a complex decrement $\lambda = \lambda_r + i\lambda_i$:

$$(\tilde{v}_1(z), \tilde{p}_1(z), \tilde{T}_1(z), \tilde{v}_2(z), \tilde{p}_2(z), \tilde{T}_2(z), \tilde{h})\exp(ik_x x + ik_y y - \lambda t); \quad (2.13)$$

where henceforwards the sign "tilde" will be omitted.

Since the problem is isotropic, similarly to the case of homogeneous fluid convection, the decrement λ depends only on the wave vector modulus $k = |k|$ but not on its direction. That is why we can limit the resolution to consideration of disturbances with $k = (k, 0)$ which do not depend on coordinate y [48]. It follows from equations (2.2) and boundary conditions (2.5), (2.6), (2.8), (2.9) that for such disturbances: the velocity components $v_{1,y}$ and $v_{2,y}$ are equal to zero, so that the motion is two-dimensional.

As a result, we get the following set of equations:

$$-\lambda v_{m,z} = -e_m p'_m + c_m D v_{m,z} + b_m G T_m,$$

$$-\lambda v_{m,x} = -ike_m p_m + c_m D v_{m,x},$$

$$-\lambda T_m + A_m v_{m,z} = \frac{d_m}{P} D T_m,$$

$$ikv_{m,x} + v'_{m,z} = 0; \quad (m = 1, 2),$$

here $D = d^2/dz^2 - k^2$ and the prime upper index denotes differentiation with respect to coordinate z. Boundary conditions are:

$$z = 1: \quad v_1 = 0, \quad T_1 = 0,$$

$$z = -a: \quad v_2 = 0, \quad T_2 = 0,$$

$$z = 0 : p_1 - p_2 + \left[Ga \left((\rho^{-1} - 1) + \delta_\beta \frac{s(1 - \rho^{-1}\beta^{-1})}{1 + \kappa a} \right) + Wk^2 \right] h =$$
$$2(v'_{1,z} - \eta^{-1} v'_{2,z}),$$

$$\eta(v'_{1,x} + ikv_{1,z}) - (v'_{2,x} + ikv_{2,z}) - ikMr\left(T_1 - \frac{s}{1 + \kappa a} h \right),$$

$$v_{1,x} = v_{2,x},$$

$$-\lambda h = v_{1,z} = v_{2,z},$$

$$T_1 - T_2 = \frac{s(1-\kappa)}{1+\kappa a} h,$$

$$\kappa T_1' - T_2' = 0.$$

Introducing the stream functions:

$$v_{m,x} = \psi_m', \qquad v_{m,z} = -ik\psi_m \qquad (m = 1, 2)$$

and eliminating pressure disturbances in a classical way, we write the boundary problem in the form:

$$\lambda D\psi_m = -c_m D^2 \psi_m + ikGb_m T_m, \tag{2.14}$$

$$-\lambda T_m - ikA_m \psi_m = \frac{d_m}{P} DT_m, \tag{2.15}$$

$$z = 1 : \psi_1 = \psi_1' = T_1 = 0, \tag{2.16}$$

$$z = -a : \psi_2 = \psi_2' = T_2 = 0, \tag{2.17}$$

$$z = 0 : \psi_1''' - \frac{1}{\eta}\psi_2''' + \left[\lambda\left(1 - \frac{1}{\rho}\right) - 3k^2\left(1 - \frac{1}{\eta}\right)\right]\psi_1' +$$
$$ik\left[Ga\left((\rho^{-1}-1) + \delta_\beta \frac{s(1 - \rho^{-1}\beta^{-1})}{1+\kappa a}\right) + Wk^2\right]h = 0, \tag{2.18}$$

$$\eta(\psi_1'' + k^2\psi_1) - (\psi_2'' + k^2\psi_2) - ikMr\left(T_1 - \frac{s}{1+\kappa a}h\right) = 0, \tag{2.19}$$

$$\psi_1' = \psi_2', \tag{2.20}$$

$$\psi_1 = \psi_2 = -i\frac{\lambda}{k}h, \qquad (2.21)$$

$$T_1 - T_2 = \frac{s(1-\kappa)}{1+\kappa a}h \qquad (2.22)$$

$$\kappa T_1' - T_2' = 0. \qquad (2.23)$$

The boundary problem (2.14)–(2.23) defines the spectrum of disturbances decrements $\lambda(k)$ as a function of eleven parameters: G, P, Mr, Ga, W, η, ν, κ, χ, β, a.

It is known that if $Ga \neq 0$, $\rho > 1$ (the upper fluid is heavier than the lower one), the Rayleigh-Taylor instability for a two-layer system takes place [92]. In the present chapter this instability is excluded since $\rho < 1$.

As mentioned in Chapter 1, the value of $Ga = G/\delta_\beta$ is large if we take into account thermogravitational convection ($G \neq 0$). It follows from (2.18) that $h = O(\delta_\beta) \ll 1$, and, within the limits of the Boussinesq approximation we assume that $h = 0$. The case of fluids of near-equal density ($|\rho^{-1} - 1| \sim \delta_\beta$) for which the interface deformation turns out to be essential is an exception. If $h = 0$ the system of boundary conditions on the interface is simplified:

$$z = 0: \quad \eta\psi_1'' - \psi_2'' - ikMrT_1 = 0, \qquad (2.24)$$
$$\psi_1' = \psi_2', \qquad (2.25)$$
$$\psi_1 = \psi_2 = 0, \qquad (2.26)$$
$$T_1 = T_2, \qquad (2.27)$$
$$\kappa T_1' = T_2'. \qquad (2.28)$$

In this case the number of parameters defining $\lambda(k)$ is reduced to nine ($G, P, Mr, \eta, \nu, \kappa, \chi, \beta, a$).

Relative surface tension change stipulated by temperature heterogeneity $\delta_\alpha = Mr/\eta W$ in many cases is not large. Then, we have parameter $W \gg 1$ for finite Mr values. In this case, for disturbances with wave numbers $k \sim 1$ we also may assume that $h = 0$ and use boundary conditions (2.24)–(2.28). The problem of the interface deformation influence on thermocapillary instability is discussed in detail in Section 2.7.

2. Non self-adjoint problem. The instability problem of a homogeneous fluids layer heated from below is known to be self-adjoint [48]. As a result, normal disturbances decrements are real-valued and the marginal curve is non oscillating.

For a two-layer system, the situation is quite different. Let us limit ourselves to the case when the interface is flat and the problem is described by equations (2.14), (2.15) with boundary conditions (2.16), (2.17), (2.24)–(2.28).

Consider a system heated from below $(s > 0; A_1 < 0, A_2 < 0)$. Using variables

$$\varphi_m = -i\sqrt{-\frac{A_m}{Gb_m}}\,\psi_m, \quad m = 1, 2$$

we rewrite equations in a symmetric form

$$c_m D^2 \varphi_m + k\sqrt{-GA_m b_m}\,T_m = -\lambda D\varphi_m,$$
$$-\frac{d_m}{P} DT_m + k\sqrt{-GA_m b_m}\,\varphi_m = \lambda T_m. \tag{2.29}$$

with boundary conditions

$$\begin{aligned} z = 1: \quad & \varphi_1 = \varphi_1' = T_1 = 0, \\ z = -a: \quad & \varphi_2 = \varphi_2' = T_2 = 0. \end{aligned} \tag{2.30}$$

$$z = 0: \quad \varphi_1'' = \varphi_2'' \frac{1}{\eta\sqrt{\kappa\beta}} + \frac{kMr}{\eta\sqrt{G(1+\kappa a)}} T_1,$$

$$\varphi_1' = \varphi_2' \frac{1}{\sqrt{\kappa\beta}}, \quad \varphi_1 = \varphi_2 = 0,$$

$$T_1 = T_2 = 0, \quad \kappa T_1' = T_2'.$$

Let us formulate a self-adjoint problem. It has the form closest to (2.29), (2.30) if we define the scalar two-component functions product $U^{(1)} = \left(\varphi_m^{(1)}, T_m^{(1)}\right)$ and $U^{(2)} = \left(\varphi_m^{(2)}, T_m^{(2)}\right)$ in the following way:

$$\left(U^{(2)}, U^{(1)}\right) = \int_0^1 dz \left(\varphi_1^{(2)*}\varphi_1^{(1)} + T_1^{(2)*}T_1^{(1)}\right) +$$

$$+ \frac{\chi}{\kappa}\int_{-a}^0 dz \left(\varphi_2^{(2)*}\varphi_2^{(1)} + T_2^{(2)*}T_2^{(1)}\right)$$

(the asterisk denotes the complex-adjoint problem). The explicit form of the adjoint problem is

$$c_m D^2 \varphi_m^c + k\sqrt{-GA_m b_m} T_m^c = -\lambda^c D\varphi_m^c,$$

$$-\frac{d_m}{P} DT_m^c + k\sqrt{-GA_m b_m}\varphi_m^c = \lambda^c T_m^c, \qquad (2.31)$$

$$z = 1 : \varphi_1^c = \varphi_1^{c'} = T_1^c = 0;$$

$$z = -a : \varphi_2^c = \varphi_2^{c'} = T_2^c = 0;$$

$$z = 0 : \varphi_1^{c''} = \varphi_2^{c''} \sqrt{\frac{\beta}{\kappa}\frac{\chi}{\nu}},$$

$$\varphi_1^{c'} = \varphi_2^{c'} \rho\chi\sqrt{\frac{\beta}{\kappa}}, \qquad (2.32)$$

$$\varphi_1^c = \varphi_2^c = 0,$$

$$T_1^c = T_2^c,$$

$$\kappa T_1^{c'} = T_2^{c'} - \varphi_2^{c'} kMr\chi P\sqrt{\frac{(1+\kappa a)\beta}{\kappa G}}.$$

If we compare boundary problems (2.29), (2.30) and (2.31), (2.32) we see that they coincide if the following conditions are realized:

$$\rho\beta\chi = 1, \quad kMr = 0. \qquad (2.33)$$

Thus, in general case the problem for the system with undeformable interface is non self-adjoint. That is why the marginal stability for the system under consideration may be both of monotonous and oscillatory character in contrast to a homogeneous fluid. In the first case the stability boundary is defined by $\lambda = 0$, and in the second by $\lambda_r = 0$ only (in this case $\lambda_i \neq 0$).

We have mentioned already that disturbances increments spectrum which is found via the boundary problem solution (2.14)–(2.17), (2.24)–(2.28) depends on nine parameters. The number of parameters is reduced, if we define the monotonous equilibrium instability boundary. Actually $\lambda = 0$ variables $T_m(z) = P\theta_m(z)$ substitution brings equations (2.14), (2.15) to the form:

$$D^2\psi_1 - ikGP\theta_1 = 0, \quad D\theta_1 + ikA_1\psi_1 = 0;$$

$$D^2\psi_2 - ikGP\frac{\nu}{\beta}\theta_2 = 0, \quad \frac{1}{\chi}D\theta_2 + ikA_2\psi_2 = 0.$$

Parameters G, P, ν, β only enter equations in the following combinations: the Rayleigh number

$$R = GP \tag{2.34}$$

and the parameter

$$\zeta = \nu/\beta \tag{2.35}$$

So, monotonous marginal stability boundary is only defined by seven independent parameters. It is worth noting that parameters R and ζ are not included in the boundary conditions.

The problem for normal disturbances is non self-adjoint even for a homogeneous fluid heated from above. In contrast to a homogeneous fluid which is always stable for heating from above, the same situation for a two-layer system might lead to instability described by surface modes (see Section 2.3).

2.2.   Thermogravitational convection when heating from below

In this section we deal with convection in a system with two horizontal layers heated from below, the thermocapillary effect ($G \neq 0, Mr = 0$) being absent. It has been mentioned in the previous section that when calculating thermogravitational convection excitation threshold, interface is considered to be undeformable ($h = 0$). In this case the boundary problem has the form:

$$c_m D^2 \psi_m - ikGb_m T_m = -\lambda D\psi_m,$$
$$\frac{d_m}{P} DT_m + ik A_m \psi_m = -\lambda T_m, \tag{2.36}$$

$$
\begin{aligned}
z = 1: &\quad \psi_1 = \psi_1' = T_1 = 0, \\
z = -a: &\quad \psi_2 = \psi_2' = T_2 = 0, \\
z = 0: &\quad \eta\psi_1'' = \psi_2'', \quad \psi_1' = \psi_2', \quad \psi_1 = \psi_2 = 0, \\
&\quad T_1 = T_2, \quad \kappa T_1' = T_2',
\end{aligned} \tag{2.37}
$$

where $b_1 = c_1 = d_1 = 1$, $b_2 = 1/\beta$, $c_2 = 1/\nu$, $d_2 = 1/\chi$, $A_1 = -1/(1 + \kappa a)$, $A_2 = -\kappa/(1 + \kappa a)$.

Let us recall (see Subsect. 2, Section 2.1) that for the case of the monotonous instability the number of problem characteristic parameters reduces to two; it is natural to choose the Rayleigh number $R = GP$ and $\zeta = \nu/\beta$ as independent parameters.

The physical mechanism of convection generation, due to unstable density stratification of the fluid is essentially similar to the case of a homogeneous fluid (see [48], Section 3). The boundary problem (2.36), (2.37) defines an infinite set of neutral curves $\lambda_r(G, k) = 0$ corresponding to disturbances which are different functions of the vertical coordinate.

Let us note an important symmetry property of the problem (2.36), (2.37). The temperature dimensional critical gradient is not changed, the upper and lower layers being formally reverted[1]. If we take into account that transition to dimensionless variables is made with respect to the upper fluid parameters, we get the following symmetry property

$$G(\eta^{-1}, \nu^{-1}, \kappa^{-1}\chi^{-1}, \beta^{-1}, P\chi/\nu, a^{-1}) = G(\eta, \nu, \kappa, \chi, \beta, P, a) \cdot \frac{\nu^2 a^3}{\beta}.$$

$$(2.38)$$

The Rayleigh number is the parameter defining convection unstability conditions for the case of a homogeneous fluid. In the system with an interface each of the media is characterised by its own (*"local"*) Rayleigh number

$$R_m = \frac{g\beta_m \tilde{A}_m a_m^4}{\nu_m \chi_m}, \quad m = 1, 2,$$

where $\tilde{A}_m$ is a dimensional temperature gradient in the mth fluid. Values of R_m are not independent and are expressed via dimensionless parameters introduced in Chapter 1 in the following way:

$$R_1 = \frac{GP}{1 + \kappa a}, \quad R_2 = \frac{GP\kappa}{1 + \kappa a} \frac{\nu\chi a^4}{\beta}.$$

The relation

$$\frac{R_2}{R_1} = \frac{\kappa\nu\chi a^4}{\beta} \tag{2.39}$$

only depends on physical properties of both media and on the value a and retaines a constant value when heating intensity is changed.

---

[1] Actually, layers cannot be reverted because of Rayleigh-Taylor instability.

Let the local Rayleigh number in one of the fluids (to be precise —
in the second) be much larger than that in the other fluid.

This case can take place if the thickness of the second layer is much
larger than that one on the first layer, for example ($a \gg 1$). Then,
naturally, on the lower fluid the convection develops at smaller G values
than in the upper fluid.

For this reason the lowest neutral curve will correspond to disturbances localized mainly in the second fluid; only a weak flow induced
by tangential stresses and temperature distribution heterogeneity on the
interface arises in the first fluid. The neutral curves characterizing convection threshold in the upper fluid lie in the region of larger Rayleigh
numbers. In limiting cases $a \ll 1$ and $a \gg 1$ the stability criteria can
be written respectively as $R_1 = R_c$ and $R_2 = R_c$, where $R_c \simeq 1708$
is the critical Rayleigh number for a layer of homogeneous fluid placed
between solid isothermal boundaries.

Stability boundaries at thermogravitational convection for infinite
horizontal layers of immiscible fluids (kerosene-sodium chloride solution system) has been calculated in the paper [12]. Constructed neutral curves related to onset of convection only in one of the media (in
kerosene). It is worth mentioning that the calculated Rayleigh number threshold value does not satisfy the finite correlation $R_1 \to R_c$ at
$a \ll 1$. It makes calculations dubious.

If local Rayleigh numbers R_1 and R_2 differ slightly, then the first
place is occupied by heat and hydrodynamic interaction on the interface. It affects the convection instability in each layer considerably.
Such a case has been considered for the first time in [26] devoted to
onset of convection in a two-layer system placed between free isothermal boundaries, the layers' thicknesses being greatly different. In this
paper two monotonous neutral curves the minima of which lie at essentially different wave numbers have been constructed for a model system
$\beta = \chi = 1; \eta = \nu = 0.1; \kappa = 0.04; a = 4$. These curves correspond to
onset of convection in each of the layers.

More interesting results were obtained in [20, 54]. It turned out that
a two-layer system has a new qualitative feature: in contrast to a homogeneous liquid, it may become convectively unstable with respect to
oscillatory disturbances. Calculations have been carried out for transformer oil-formic acid system (see Table 1). The Rayleigh numbers ratio (2.39) for this system is $6.71a^4$. Calculation results derived by the
Runge-Kutta method are presented in Fig. 2.1–2.4.

Fig. 2.1 presents two neutral curves, constructed for $a \simeq 0.818$
($R_2/R_1 = 3.01$); eigenfunctions at points A and B are given in Fig. 2.2.
The disturbance on the lower neutral curve is concentrated in the sec-

Table 1   Parameters of Two-layer Systems

System	P	η	ν	κ	χ	β
Water–Silicone Oil DC 200	6.28	0.915	1.116	0.169	0.472	7.1595
Air–Water	0.758	0.0182	15.1	0.0396	138	17.7
Transformer Oil Formic Acid	306	11.1	15.4	0.41	0.714	0.672
Olive Oil 85% Glycerine Solution	970	0.74	0.985	0.54	0.88	1.48

ond fluid mainly (formic acid) characterized by a higher local Rayleigh number value.

Now let us consider the case of $a = 0.667$ for which local Rayleigh numbers are close: $R_2/R_1 = 1.33$. In this case an original "crossing" of the neutral curves 1 and 2 takes place corresponding to convection arising in the lower and upper fluids (see Fig. 2.3). In the crossing region the monotonous neutral curves closing takes place and the limit of oscillatory marginal instability is reached. Let us emphasize that monotonous instability related to convection in the lower fluid (neutral curve 2 minimum) is held to be the most dangerous one.

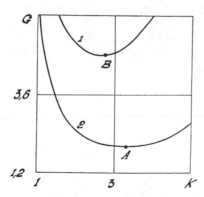

Figure 2.1. The neutral curves for a transformer oil-formic acid system; line 1 — onset of convection in the upper fluid, line 2 — in the lower fluid.

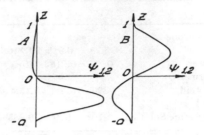

Figure 2.2. The form of neutral disturbances relative to points A and B of Figure 2.1.

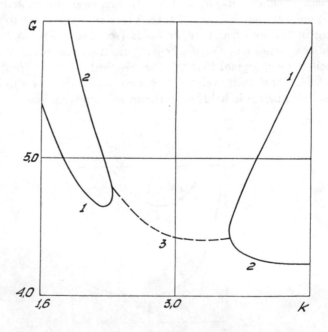

Figure 2.3. The neutral curves at $a=0.667$. Solid lines — monotonous instability boundaries, dash line — oscillatory instability boundary.

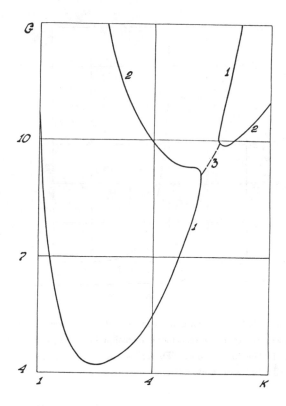

Figure 2.4. The neutral curves at $a=0.54$. Lines 1, 2 — monotonous instability boundaries, line 3 — oscillatory instability boundary.

Further decrease of a results in that the minimum of neutral curve 1 corresponding to the upper fluid disturbances becomes lower than curve 2 minimum. That is why convection develops in the oil layer first (see Fig. 2.4; $R_2/R_1 = 0.57$).

So, for transformer oil-formic acid system the convection threshold is connected with monotonously increasing disturbances localized in the lower fluid at $a > a_*$ and in the upper fluid at $a < a_*$, where $a_* \simeq 0.6$.

The paper [2] presents an experimental study of convection generation in transformer oil-formic acid system. According to the experiment conditions, the total layer thickness $a_1 + a_2$ was defined by the cavity

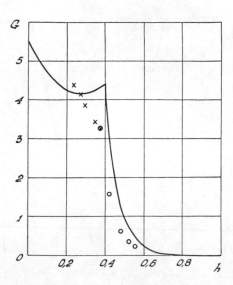

Figure 2.5. Convection generation in a transformer oil-formic acid system. Crosses relate to convection generation in the upper fluid, circles — to that in the lower fluid. (Experimental data of paper [2]. Solid line — calculation results of [54]).

geometry and was constant, while the ratio $h = a_2/(a_1 + a_2)$ varied. The authors assumed that thermocapillary effect was insignificant. The convection threshold was determined according to the heat-flow curve break. Experimental results and theoretical curves are given in Fig. 2.5. In the region of $h < h_* \simeq 0.37$ convection arose in the transformer oil layer (the upper fluid), and at $h > h_*$ it arose in the formic acid layer (the lower fluid). When $h = h_*$, convection arose in both fluids simultaneously. The value of h_* obtained experimentally agrees with theoretical prediction.

The problem of oscillatory instability, which is the most dangerous one for some real fluid systems, remains still unsolved today. It has been found that formal variation of the problem parameters (2.36)–(2.37) can lead to the case where the minimum Grashof number corresponds to oscillatory disturbances. Let us quote an interesting observation [59]. If we take the previous system and reduce parameter η, then the oscillation

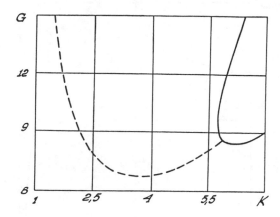

Figure 2.6. The neutral curves for the model system (a=0.54).

region increases considerably. For the case given in Fig. 2.6–2.7 ($\eta =$ 0.123) the given region occupies the interval $|k| < k_m \simeq 5.8$. The neutral curve minimum is realized at $k = k_c \simeq 3.8$ for oscillatory disturbances with frequency $\omega_c \simeq 0.051$.

The model system studied earlier in [26] is again considered in a recent paper [136]. As ought to be expected, an oscillatory instability region in the interval between two monotonous neutral curves was discovered. The same paper quotes calculations of neutral curves for a series of other model systems in some interval of layers with an oscillatory instability thickness ratio. An example proving that oscillatory instability is the most dangerous one is given.

2.3. Instability when heating from above

Now let us consider the case of thermogravitational convection ($Mr =$ 0) when heating from above. The steady state is described by the boundary problem (2.36), (2.37); the temperature gradients are positive: $A_1 = 1/(1 + \kappa a)$, $A_2 = \kappa/(1 + \kappa a)$.

It is known that homogeneous fluid steady state is stable when the temperature gradient is directed upwards. Instability when heating from

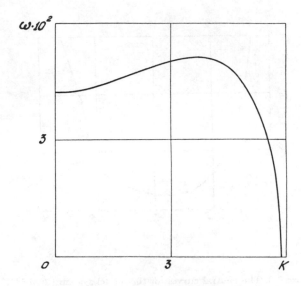

Figure 2.7. The frequency of neutral oscillatory disturbances dependence on the wave number.

above, due essentially to heat and hydrodynamic interaction on the interface becomes possible. This is a specific feature of a two-layer system.

1. Model system. Physical mechanism of instability can be explained using the example of a system in which the lower fluid temperature diffusivity is much higher than that of the upper one ($\chi \ll 1$), and the upper fluid thermal expansion is relatively small ($\beta \ll 1$) [49].

Let an element of the upper fluid move down towards the interface, the movement being caused by an accidental disturbance. Owing to low temperatre diffusivity of the upper fluid, the element temperature remains higher than that of the surrounding fluid for a long time. But it does not make the element rise to the surface because the upper fluid thermal expansion is negligibly small. The temperature field change on the interface causes free convective motion in the lower fluid shown in Fig. 2.8. Velocity and tangential stress continuity on the interface causes a flow induction in the upper fluid intensifying the initial disturbance. Similar considerations can be applied to the case when the upper fluid temperature diffusivity is high ($\chi \gg 1$) and heat expansion coefficient of the lower fluid is small ($\beta \gg 1$).

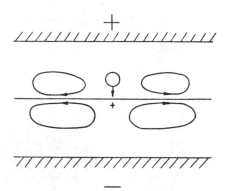

Figure 2.8. For explanation of instability mechanism when heating from above.

In the limit $\chi_2 \to \infty, \beta_1 = 0$ the marginal stability boundary can be calculated analytically[2]. Since the Grashof number G determined according to the upper fluid parameters vanishes as $\beta_1 = 0$, at this point the Grashof number G' determined according to the lower fluid parameters will be used:

$$G' = \frac{g\beta_2 \theta a_2^3}{\nu_2^2}.$$

In this limit, equations (2.36) for $\lambda = 0$ (stability boundary) after replacement of variables are reduced to the form:

$$D^2\psi_1 = 0, \quad D\theta_1 + ikA_1\psi_1 = 0,$$
$$D^2\psi_2 - ikR'\theta_2 = 0, \quad D\theta_2 = 0,$$

where R' is the crucial parameter (effective Rayleigh number):

$$R' = \frac{G'P}{\nu a^3}. \tag{2.40}$$

---

[2] Typical Rayleigh convection when heating from below is impossible in the system considered, because there is no thermal heat expansion in the upper fluid and there is no temperature gradient in the lower one.

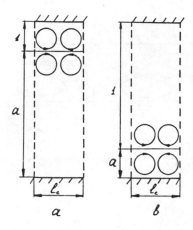

Figure 2.9. The structure of critical disturbances $a)a \gg 1$; $b)a \ll 1$.

Boundary conditions are similar to (2.37). The boundary problem determines the neutral curve $R'(k)$:

$$R'(k) = \frac{32(1 + \kappa a)k^4}{\kappa}$$
$$\frac{(\kappa S_2 C_1 + S_1 C_2)[\eta(S_1 C_1 - k)(S_2^2 - k^2 a^2) + (S_2 C_2 - ka)(S_1^2 - k^2)]}{(S_2^3 - k^3 a^3 C_2)(S_1^3 - k^3 C_1)},$$

$$(2.41)$$

where $S_1 = \mathrm{sh}k$, $C_1 = \mathrm{ch}k$, $S_2 = \mathrm{sh}ka$, $C_2 = \mathrm{ch}ka$. Asymptotically,

$$R'(k) \simeq \frac{1600}{k^2} \frac{(1 + \kappa a)^2(1 + \eta a)}{\kappa a^4} \quad (k \to 0);$$

$$R'(k) \simeq \frac{32(1 + \kappa a)(1 + \kappa)(1 + \eta)}{\kappa} k^4 \quad (k \to \infty).$$

The wave number k_c of the most dangerous disturbance corresponding to the function minimum $R'(k)$ when $a = 1$ does not depend on κ and η and equals 1.58 [49]. When $a \neq 1$ k_c depends on all parameters. Observe that when layer thicknesses differ greatly, the critical wave length $l_c = 2\pi/k_c$ is of more thin layer thickness type, i.e. $k_c \sim 1$ as $a \gg 1$, $k_c \sim a^{-1}$ as $a \ll 1$ (see Fig. 2.9 a,b).

Formula (2.41) determines instability threshold when $\chi \ll 1, \beta \ll 1$. In the opposite case $\chi \gg 1, \beta \gg 1$, the critical Grashof number can be calculated according to (2.38).

2. Shortwave asymptotic of neutral curves. In the general case to find out whether instability is possible when heating is from above, the problem (2.36), (2.37) must to be solved numerically. However, the conditions of instability within a shortwave limit can be obtained analytically [187]. Introducing variable

$$Z = kz$$

and denoting

$$\vartheta_m = kT_m, \quad G = k^4 \mathcal{G}$$

we rewrite equations (2.36) without the parameter k:

$$c_m(\psi_m^{IV} - 2\psi_m'' + \psi_m) - i\mathcal{G}b_m\vartheta_m = 0,$$
$$\frac{d_m}{P}(\vartheta_m'' - \vartheta_m) + iA_m\psi_m = 0. \tag{2.42}$$

(prime denotes differentiation with respect to Z). Boundary conditions take the form

$$Z = k : \psi_1 = \psi_1' = \vartheta_1 = 0; \quad Z = -ka : \psi_2 = \psi_2' = \vartheta_2 = 0, \tag{2.43}$$

$$Z = 0 : \eta\psi_1'' = \psi_2'', \psi_1' = \psi_2', \psi_1 = \psi_2 = 0, \vartheta_1 = \vartheta_2, \kappa\vartheta_1' = \vartheta_2'. \tag{2.44}$$

When $k \to \infty$ the problem can be simplified because shortwave disturbances localize in a narrow layer $|z| \sim l(|Z| \sim 1)$ close to the interface (see Fig. 2.10). Owing to this, boundary conditions (2.43) can be expressed in the following way:

$$Z \to \infty : \psi_1, \vartheta_1 \to 0; \quad Z \to -\infty : \psi_2, \vartheta_2 \to 0. \tag{2.45}$$

The solution of equations (2.42) satisfying conditions (2.45) has the form

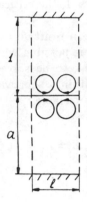

Figure 2.10. The structure of shortwave disturbances.

$$\psi_1 = \sum_{i=1}^{3} C_i^{(1)} e^{-q_i^{(1)} z}, \quad \theta_1 = \sum_{i=1}^{3} D_i^{(1)} e^{-q_i^{(1)} z},$$

$$\psi_2 = \sum_{i=1}^{3} C_i^{(2)} e^{q_i^{(2)} z}, \quad \theta_2 = \sum_{i=1}^{3} D_i^{(2)} e^{q_i^{(2)} z}. \tag{2.46}$$

The values of $q_i^{(m)} (m = 1, 2)$ are determined by conditions

$$q_i^{(m)2} = 1 + \alpha_i R_m \mathcal{G}^{1/3}, \quad \mathrm{Re} q_i^{(m)} > 0,$$

$$\alpha_1 = 1, \quad \alpha_{2,3} = -\frac{1}{2} \pm \frac{\sqrt{3}}{2} i, \quad R_m = \left(P \frac{A_m b_m}{c_m d_m} \right)^{1/3}.$$

The constants $C_i^{(m)}, D_i^{(m)}$ are linked by equations

$$D_i^{(m)} = -i \mathcal{G}^{-1/3} Q_m \alpha_i^{-1} C_i^{(m)}, \tag{2.47}$$

where $Q_m = (A_m^2 P^2 c_m / d_m^2 b_m)^{1/3}$.

Substituting solution (2.46) into boundary conditions (2.44) and taking (2.47) into consideration, we get a system of six homogeneous linear algebraic equations in the coefficients $C_i^{(m)} (i = 1, 2, 3; m = 1, 2)$. Setting the determinant of the system equal to zero,

$$
\begin{vmatrix}
1 & 1 & 1 & 0 & 0 & 0 \\
0 & 0 & 0 & -1 & -1 & -1 \\
-q_1^{(1)} & -q_2^{(1)} & -q_3^{(1)} & -q_1^{(2)} & -q_2^{(2)} & -q_3^{(2)} \\
\eta q_1^{(1)^2} & \eta q_2^{(1)^2} & \eta q_3^{(1)^2} & -q_1^{(2)^2} & -q_2^{(2)^2} & -q_3^{(1)^2} \\
\dfrac{Q_1}{\alpha_1} & \dfrac{Q_1}{\alpha_2} & \dfrac{Q_1}{\alpha_3} & \dfrac{-Q_2}{\alpha_1} & \dfrac{-Q_2}{\alpha_2} & \dfrac{-Q_2}{\alpha_3} \\
\dfrac{-\kappa q_1^{(1)} Q_1}{\alpha_1} & \dfrac{-\kappa q_2^{(1)} Q_1}{\alpha_2} & \dfrac{-\kappa q_3^{(1)} Q_1}{\alpha_3} & \dfrac{-q_1^{(2)} Q_2}{\alpha_1} & \dfrac{-q_2^{(2)} Q_2}{\alpha_2} & \dfrac{-q_3^{(2)} Q_2}{\alpha_3}
\end{vmatrix} = 0
$$

$$(2.48)$$

we get a transcendent equation determining the value $\mathcal{G}$ as a function of the fluids' parameters. In general, this equation cannot be solved analytically. But there is no need of doing this when the stability criterion is determined. It is quite sufficient to find the boundary in the space of parameters dividing it into two regions: the region where exists a solution corresponding thus to an instability region and the region where the equation has no material solutions (stability region). The critical Grashof number is naturally expected to go to infinity when we approach the boundary.

In the limit $\mathcal{G} \to \infty$, (2.48) can be brought to the form

$$
\begin{vmatrix}
1 & 1 & 1 & 0 & 0 & 0 \\
0 & 0 & 0 & -1 & -1 & -1 \\
-1 & \alpha_3 & \alpha_2 & -R & \alpha_3 R & \alpha_2 R \\
1 & \alpha_2 & \alpha_3 & -\eta^{-1} R^2 & -\eta^{-1}\alpha_2 R^2 & -\eta^{-1}\alpha_3 R^2 \\
1 & \alpha_3 & \alpha_2 & -Q & -\alpha_3 Q & -\alpha_2 Q \\
-1 & \alpha_2 & \alpha_3 & -\kappa^{-1} RQ & \kappa^{-1}\alpha_2 RQ & \kappa^{-1}\alpha_3 RQ
\end{vmatrix} = 0
$$

where

$$
R = \left(\frac{R_2}{R_1} \right)^{1/2} = (\kappa \zeta \chi)^{1/6},
$$

$$
Q = \frac{Q_2}{Q_1} = (\kappa \chi)^{2/3} \zeta^{-1/3},
$$

$$(2.49)$$

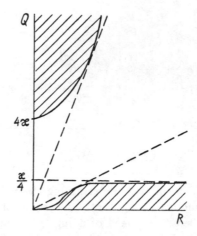

Figure 2.11. The regions of the shortwave instability existence when heating from above (hatched).

$$\left(\zeta = \frac{\nu}{\beta}\right).$$

Developping the determinant, we get the equation of shortwave instability boundary in the space of the system parameters:

$$F = \eta Q^2 - [3(\kappa + \eta)R + 4\kappa\eta + 4R^2]Q + \kappa R^2 = 0. \qquad (2.50)$$

A shortwave instability obviously occurs when $F > 0$; when $F < 0$ the steady state turns out to be stable with respect to shortwave disturbances. The forms of the curves determined by equation (2.50) are shown qualitatively in Fig. 2.11 (solid lines). Instability regions (hatched area in Fig. 2.11) lie outside sector $\gamma_1 < Q/R < \gamma_2$, whose the boundaries are shown by dashed lines. In this case, the values of γ_1, γ_2 only depend on the parameter $\xi = (\kappa/\eta)^{1/2}$:

$$\gamma_{1,2} = \frac{1}{2}\left[3\xi^2 + 8\xi + 3 \pm \sqrt{(3\xi^2 + 8\xi + 3)^2 - 4\xi^2}\right].$$

Let us observe that $\gamma_1 < 1/3, \gamma_2 > 3$ for any ξ; that is why the fluid systems for which the ratio Q/R lies within the interval $(1/3, 3)$ is

Table 2   Water-Mercury System

$t,^\circ\ C$	ζ	κ	χ	η
10	22.3	0.0685	0.0311	0.806
20	7.69	0.0694	0.0314	0.645

definitely stable with respect to shortwave disturbances. Besides, from inequality $\gamma_1 < \frac{1}{3}\xi^2, \gamma_2 > 3\xi^2$ it follows that the system is stable if the parameter $Q\eta/\kappa R$ lies in the interval $(1/3,3)$ (this criterion is quoted in [187]).

3. Numerical results. The criterion $F > 0$ obtained from the analysis of a neutral curve shortwave ($k \to \infty$) asymptotics is not a necessary condition of equilibrium instability when heating from above. In fact, this instability can take place in a finite interval of the wave numbers. As an example, let us consider water-mercury system calculated in [49]. Table 2 presents the system parameters at 10° and 20°. Substituting values of F we find that at 10° C shortwave equilibrium instability ($F > 0$) takes place, and at 20° C there is no shortwave instability ($F < 0$). But, according to numerical calculations, in the latter case the steady state turns to be unstable with respect to disturbances when wave numbers are in an interval $0 < k < k_*$. Neutral curves for both cases are given in Fig. 2.12.

Numerical calculations carried out in [49] for equal thickness layers allow us to draw certain conclusions. In finite case of large ζ and χ^{-1}, as it was mentioned in section 1 above, the critical Grashof number can be determined according to (2.40), (2.41). In this case the value of Q determined by (2.49) is small and the neutral curve is similar to curve 1 (Fig. 2.12) (a shortwave instability takes place). When ζ and χ^{-1} decrease, the value of Q grows. When the boundary determined by (2.50) is crossed, the shortwave instability vanishes. Neutral curves become sack-shaped (see line 2, Fig. 2.12): for fixed k instability region is limited in terms of G from below and from above. In the region of small k the critical G numbers on both branches change according to the law $G \sim 1/k^2$. When certain finite values of ζ_* and χ_*^{-1} are reached, the minimum critical Grashof number goes to infinity and the corresponding wave number goes to zero. If $\zeta < \zeta_*$ or $\chi^{-1} < \chi_*^{-1}$, there is no instability. The region in which the reference state for equal thickness layers when heating from above is stable at any κ and η, is hatched in Fig. 2.13.

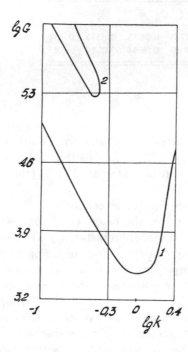

Figure 2.12. The neutral curves for a water-mercury system at $10°$C (line 1) and $20°$C (line 2); a=1.

2.4.   System with heat-insulated boundaries

The convective stability of homogeneous fluid is known to be influenced by heat conditions on solid boundaries [48]. In this paragraph the results for a two-layer system with heat-insulated outer boundaries are presented [51].

As in the previous paragraphs, equations (2.36) are used. Assuming cases of heating from below and heating from above to be considered, let $A_1 = -s/(1 + \kappa a)$, $A_2 = -s\kappa/(1 + \kappa a)$, where $s = 1$ for heating from below, $s = -1$ for heating from above. The following conditions apply at solid boundaries:

$$z = 1: \quad \psi_1 = \psi_1' = T_1' = 0,$$
$$z = -a: \quad \psi_2 = \psi_2' = T_2' = 0; \tag{2.51}$$

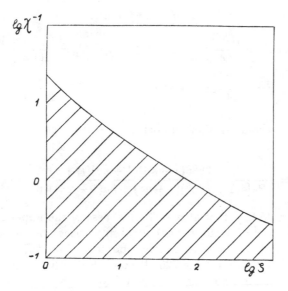

Figure 2.13. The region of stability when heating from above; $a=1$.

the conditions on the interface ($z = 0$) remain as in (2.37). Note that instability criteria when heating from above in the shortwave limit ($k \to \infty$) discussed in 2 Section 2.3 do not depend on the solid boundary conditions and remain valid. But shortwave disturbances are not the most dangerous ones; that is why investigation of the marginal stability, when the boundaries are heat-insulated, requires additional analysis.

1. Longwave instability. Longwave disturbances ($k \to 0$) are the most dangerous ones for a homogeneous fluid layer with heat-insulated boundaries. Longwave disturbances are naturally expected to play an important role also in the case of a two-layer system.

We seek solution of the problem in the form of expansion:

$$T_i = T_i^{(0)} + k^2 T_i^{(2)} + \cdots, \quad \psi_i = k\psi_i^{(1)} + k^3 \psi_i^{(3)} + \cdots,$$
$$G = G^0 + k^2 G^{(2)} + \cdots.$$

We find

$$T_1^{(0)} = T_2^{(0)} = 1;$$

$$\psi_1^{(1)} = \frac{iG^{(0)}}{24}z(z-1)^2(z+C_1), \quad C_1 = \frac{(\eta - \zeta a^2)a}{2(1+\eta a)};$$

$$\psi_2^{(1)} = \frac{iG^{(0)}}{24}\zeta z(z+a)^2(z+C_2), \quad C_2 = \frac{\eta - \zeta a^2}{2a\zeta(1+\eta a)}.$$

For the threshold Grashof number at $k \to 0$ we get:

$$G^{(0)} = s\frac{2880(1 + \kappa a)(1 + \eta a)(\kappa + a)}{P\kappa[4(1 + \chi\zeta a^5)(1 + \eta a) + 5a(1 - \chi a^2)(\eta - \zeta a^2)]}. \qquad (2.52)$$

Let us note first that at $a \to 0$ (thickness of the upper layer is much larger than that of the lower one) instability arises when heating from below, the threshold Rayleigh local number for the fluid $R_1 = GP/(1+\kappa a)$ introduced in Section 2.2 is 720. It coincides with the critical Rayleigh number value for a homogeneous fluid layer with heat-insulated boundaries [48]. Similarly, at $a \to \infty$ (the thickness of the lower layer is much larger than that of the upper one) the critical local Rayleigh number for the second fluid $R_2 = GP\kappa\nu\chi a^4/(1 + \kappa a)\beta$ has the same value.

Two types of the dependence of $G^{(0)}$ on a are possible in general. If the polynom in the denominator of expression (2.52) does not change its sign as a is changed (in particular, it takes place at $\eta\chi/\zeta = 1$) then for any a, instability only develops when heating from below. But this polynom may become negative in some interval of $a_- < a < a_+$; in this interval the longwave instability is realized when heating from above (see Fig. 2.14). Let us note that values of $a_{\pm}^2$ are contained between χ^{-1} and $\eta\zeta^{-1}$.

Limits of the above-mentioned instability for different values of η are given in Fig. 2.15. To the right and higher than the lines given in this figure, there is no instability for any a values. The results given in Fig. 2.15 relate to stability boundaries for small ζ^{-1} and χ. By virtue of symmetry (2.38), there are corresponding instability regions for large ζ^{-1} and χ.

2. Numerical results. A stability calculation for arbitrary k can be carried out on the basis of numerical methods only.

As an example let us discuss results for the water-mercury system stability at $10°C$ (the system parameters are given in Table 2) which is characterized by great difference of layers properties (small ζ^{-1} and χ). The stability boundary in relation to longwave disturbances has a

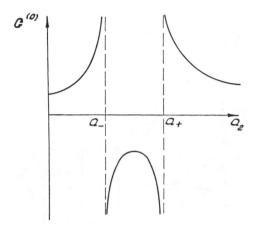

Figure 2.14. The threshold value of the Grashof number dependence on the layers thickness ratio for a longwave instability.

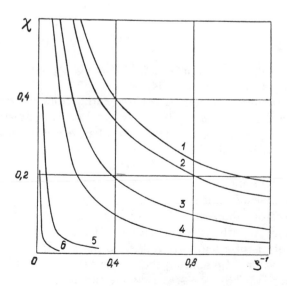

Figure 2.15. Boundaries of longwave instability existence regions as $\eta=0$ (line 1), 0.1 (2), 0.5 (3), 1 (4), 5 (5), 10 (6).

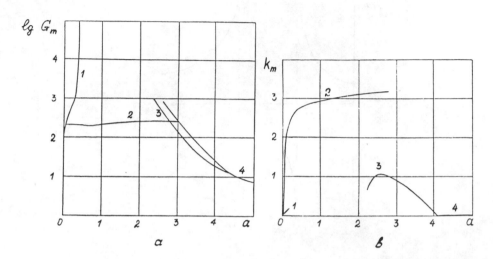

Figure 2.16. Dependences of the critical Grashof number (a) and the critical wave number (b) on the parameter a for a water-mercury system when heating from below.

form similar to the one given in Fig. 2.14; the discontinuity points are observed for $a_- = 0.397$, $a_+ = 2.94$.

Rayleigh instability characteristics when heating from below are given in Fig. 2.16. In the regions $a < 0.03$ and $a > 4.3$ longwave disturbances are the most dangerous ones (branches 1,4). Parameter a having intermediate values, stability boundary is formed by branches 2 and 3 crossing as $a = 2.6$. Branch 2 corresponds to cellular structures arising in the upper layer (water) and branch 3 to those in the lower one (mercury).

The critical Grashof number and the critical wave number as functions of a for heating from above are presented in Fig. 2.17. As $a < 0.418$ and $a > 2.66$ cellular disturbances are the most dangerous ones (branches 2, 3). Instability is of a longwave nature (branch 1) in the intermediate region of a values. Comparing the results to the case of isothermal boundaries (Section 2.3), one sees that for the case of heat-insulated boundaries, the instability threshold for heating from above is considerably lower (for $a = 1$, we get for isothermal and heat-insulated boundaries $G_m = 3.01 \cdot 10^3$ and 150 respectively).

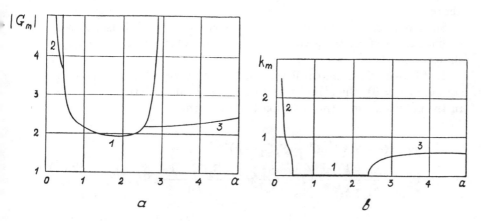

Figure 2.17. Dependences of the critical Grashof number (a) and the critical wave number (b) on parameter a for a water-mercury system when heating from above.

2.5. Thermocapillary convection

Let us consider now another mechanism of a two horizontal layer system instability connected with thermocapillary effect $(Mr \neq 0)$. We shall restrict ourselves with the limit $G \to 0$ (convection in thin films).

As in 1 Section 2.1, when parameter $W \gg 1$, the boundary distortion caused by convection has to be taken into account for longwave disturbances only. In Sections 2.5, 2.6 we shall put $h = 0$ and use boundary conditions (2.24)–(2.28). The boundary deformation effect on stability is studied in Section 2.7.

The problem (2.14), (2.15), (2.24)–(2.28) has the form:

$$c_m D^2 \psi_m = -\lambda D \psi_m,$$
$$\frac{d_m}{P} DT_m + ik A_m \psi_m = -\lambda T_m \quad (m = 1, 2),$$

(2.53)

$$z = 1 : \psi_1 = \psi_1' = T_1 = 0; \quad z = -a : \psi_2 = \psi_2' = T_2 = 0; \quad (2.54)$$
$$z = 0 : \eta \psi_1'' - \psi_2'' - ik Mr T_1 = 0, \quad \psi_1' = \psi_2',$$

(2.55)

$$\psi_1 = \psi_2 = 0, \quad T_1 = T_2, \quad \kappa T_1' = T_2'.$$

Here $c_1 = d_1 = 1$, $c_2 = 1/\nu$, $d_2 = 1/\chi$, $A_1 = -s/(1 + \kappa a)$, $A_2 = -s\kappa/(1 + \kappa a)$; $s = 1$ for heating from below, $s = -1$ for heating from above.

Since boundary problem for disturbances at $Mr \neq 0$ is not self-adjoint, the neutral state can be of both monotonous and oscillatory character.

1. Monotonous instability. On the boundary of monotonous instability ($\lambda = 0$), the solution has been obtained analytically in [169]. In the notation used, this solution has the form:

$$\psi_1 = S - S_1^2 k^{-1} zC + (S_1 C_1 - k)k^{-1} zS,$$
$$\psi_2 = \frac{(S_1^2 - k^2)a^2}{S_2^2 - k^2 a^2}\left(S - \frac{S_2^2}{ka^2}zC - \frac{S_2 C_2 - ka}{ka^2}zS\right), \tag{2.56}$$

$$T_1 = e_1(S - t_1 C) + \frac{isP}{4k^2(1 + \kappa a)}[-k(2t_1 + k)C-$$
$$-(S_1 C_1 - k)zS + (S_1^2 + 2k^2)zC - kS_1^2 z^2 S + k(S_1 C_1 - k)z^2 C],$$
$$T_2 = e_2(S + t_2 C) + \frac{isP\chi\kappa}{4k^2(1 + \kappa a)}\frac{S_1^2 - k^2}{S_2^2 - k^2 a^2} \times$$
$$\times[ka^2(2t_2 + ka)C + (S_2 C_2 - ka)zS + (S_2^2 + 2k^2 a^2)zC-$$
$$-kS_2^2 z^2 S - k(S_2 C_2 - ka)z^2 C],$$

where

$$S = \mathrm{sh}\,kz, \quad C = \mathrm{ch}\,kz, \quad S_1 = \mathrm{sh}\,k, \quad C_1 = \mathrm{ch}\,k,$$
$$t_1 = \mathrm{th}\,k, \quad S_2 = \mathrm{sh}\,ka, \quad C_2 = \mathrm{ch}\,ka, \quad t_2 = \mathrm{th}\,ka,$$

$$e_1 = -isP\{(S_2^2 - k^2 a^2)[(2t_1 + k)k^2 + \kappa t_2(S_1^2 + 2k^2)]-$$
$$-\chi\kappa(S_1^2 - k^2)(S_2^2 t_2 - k^3 a^3)\}/[4k^3(1 + \kappa a) \cdot (t_1 + \kappa t_2)(S_2^2 - k^2 a^2)],$$
$$e_2 = isP\kappa\{(S_1^2 t_1 - k^3)(S_2^2 - k^2 a^2) - \chi(S_1^2 - k^2) \cdot [t_1(S_2^2 + 2k^2 a^2)+$$
$$+\kappa k^2 a^2(2t_2 + ka)]\}/[4k^3(1 + \kappa a)(t_1 + \kappa t_2)(S_2^2 - k^2 a^2)].$$

The neutral curve is determined from

$$Mr = Mr_m = 8sk^2\frac{1 + \kappa a}{P\kappa}\frac{(\kappa D_1 + D_2)(\eta B_1 + B_2)}{\chi E_2 - E_1}, \tag{2.57}$$

where

$$D_1 = \frac{1}{t_1}, D_2 = \frac{1}{t_2}, B_1 = \frac{S_1 C_1 - k}{S_1^2 - k^2}, B_2 = \frac{S_2 C_2 - ka}{S_2^2 - k^2 a^2},$$

$$E_1 = \frac{S_1^2 t_1 - k^3}{t_1(S_1^2 - k^2)}, E_2 = \frac{S_2^2 t_2 - k^3 a^3}{t_2(S_2^2 - k^2 a^2)}.$$

As mentioned in Section 2.2, the existence of an infinite number of neutral curves corresponding to different convective modes is a characteristic feature of Rayleigh instability. On the contrary, for the case of thermocapillary convection a monotonous neutral curve turns out to be unique. Neutral curve can have a discontinuity at some k value determined by the equation

$$\frac{f(ka)}{f(k)} = \chi^{-1},$$

where

$$f(x) = \frac{\text{sh}^2 x \cdot \text{th} x - x^3}{\text{th} x (\text{sh}^2 x - x^2)}.$$

The ratio $f(ka)/f(k)$ for any $a \neq 1$ is a monotonous function of parameter k changing from a^2 (as $k \to 0$) to 1 (as $k \to \infty$). This result is a consequence of the function $r(x) = x f'(x)/f(x)$ monotony and relation

$$\frac{f(ka)}{f(k)} = \exp \int_1^a r(ka') da'$$

(function $r(x)$ plot is given in Fig. 2.18). So, the neutral curve either has no discontinuities (if χ^{-1} lies outside the range $f(ka)/f(k)$), or has only one discontinuity.

Depending on the parameters χ and a values the neutral curve corresponds to one of the four types given in Fig. 2.19. If $\chi > 1$ the case a is realized (if $a > a_* = \chi^{-1/2}$) or the case b (if $a < a_*$). If $\chi < 1$ then either the case c ($a < a_*$) or the case d ($a > a_*$) takes place.

Let us give asymptotics for (2.57) in a longwave and shortwave limits:

$$k \to 0: \quad Mr = -\frac{80s(1 + \kappa a)^2(1 + \eta a)}{Pa^2 \kappa(1 - \chi a^2)} k^{-2}, \tag{2.58}$$

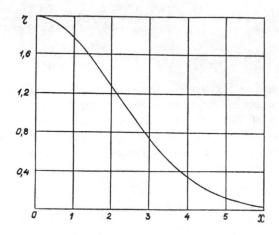

Figure 2.18. The graph of $r(x)$ function.

$$k \to \infty: \quad Mr = \frac{8s(1 + \kappa a)(1 + \kappa)(1 + \eta)}{P\kappa(\chi - 1)}k^2. \qquad (2.59)$$

Let us now discuss the case $a = 1$ for which formula (2.57) gets the form:

$$Mr = \frac{8sk^2(1 + \kappa)^2(1 + \eta)}{\kappa P(\chi - 1)}\frac{S_1 C_1 - k}{S_1^2 t_1 - k^3}.$$

It follows from this relation, that monotonous neutral curve has no discontinuities when $a = 1$. Instability arises only when heating is from the fluid side having the lower temperature diffusivity; the opposite heating method being used, the monotonous instability is impossible. Let us pay attention to the fact that when fluids have got equal temperature diffusivities ($\chi = 1$) at $a = 1$ there is no monotonous neutral curve at all.

An example of monotonous neutral curves calculated according to (2.57) for a real water-silicone oil system (see Table 1; $a_* = 1.46$) is given in Fig. 2.20 [110].

2. Oscillatory instability (analytical results). The oscillatory instability boundary for arbitrary parameters cannot be calculated analytically. But oscillatory instability existence can be determined on

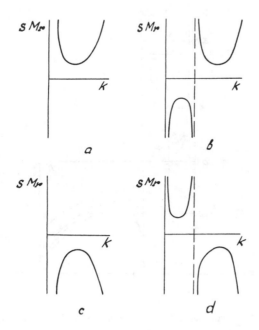

Figure 2.19. The neutral curves types for monotonous thermocapillary instability.

the basis of decrement $\lambda(Mr)$ behaviour close to the monotonous instability boundary. Actually, formula (2.57) describes the monotonous marginal stability only in the case where for $Mr < Mr_m$ there is no oscillatory instability (see Fig. 2.21a). Otherwise formula (2.57) meets the stabilization boundary of one of the unsteady monotonous modes (see Fig. 2.21b). Obviously, the sign of the oscillatory instability existence is determined from

$$\lambda' \equiv \left(\frac{d\lambda}{dMr}\right)_{Mr=Mr_m} > 0.$$

The expression for λ' can be obtained from (2.53)–(2.55) expansion of λ, ψ_m, T_m in $Mr - Mr_m$. From the problem solvability condition in the first order with respect to $Mr - Mr_m$ we find:

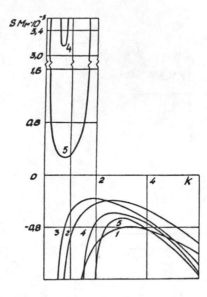

Figure 2.20. The neutral curves for a water-silicone oil No. 200 system as a=0.2 (1), 0.4 (2), 1 (3), 1.6 (4), 2.4 (5).

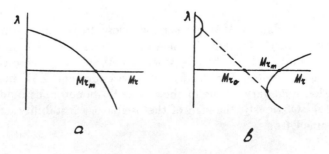

Figure 2.21. Decrement λ dependence on parameter Mr : a) the case of monotonous instability; b) the case of oscillatory instability.

$$\lambda' = -\frac{F_2}{F_1},$$

$$F_1 = \int\limits_{0}^{1} dz(\psi_1^c D\psi_1 + T_1^c T_1) + \int\limits_{-a}^{0} dz(\psi_2^c D\psi_2 + T_2^c T_2),$$

$$F_2 = \frac{\kappa P(s_1^2 - k^2)}{32k^4(1+\kappa a)^2} \frac{(E_1 - \chi E_2)^2}{(B_2 + \eta B_1)(D_2 + \kappa D_1)},$$

where ψ_1, T_1, ψ_2, T_2 are determined by (2.56)

$$\psi_1^c = \frac{s}{8k(1+\kappa a)} \left\{ \frac{\eta[(1+kD_1) - \chi(1+kaD_2)] + k(1+\chi\eta a)B_2}{(B_2 + \eta B_1)(S_1^2 - k^2)} \times \right.$$
$$\left. \times \left(S - \frac{S_1^2}{k}zC + \frac{S_1 C_1 - k}{k}zS \right) - z^2C + \frac{C_1}{S_1}z^2S + zC - \frac{C_1}{S_1}zS \right\},$$

$$\psi_2^c = \frac{s\nu}{8k(1+\kappa a)} \left\{ \frac{[(1+kD_1) - \chi(1+kaD_2) - k(1+\chi\eta a)B_1}{(B_2 + \eta B_1)(S_2^2 - k^2a^2)} \times \right.$$
$$\times \left[\left(a^2S - \frac{S_2^2}{k}zC - \frac{S_2 C_2 - ka}{k}zS \right) + \right.$$
$$\left. \left. \chi \left(-z^2C - \frac{C_2}{S_2}z^2S - azC - \frac{aC_2}{S_2}zS \right) \right] \right\},$$

$$T_1^c = i\left(C - \frac{C_1}{S_1}S \right), \quad T_2^c = \frac{i\chi}{\kappa}\left(C + \frac{C_2}{S_2}S \right).$$

Let us note that at the neutral curve discontinuity $(E_1 - \chi E_2) = 0$ F_2 does not change the sign and F_1 (as opposed to S) changes sign. As a result, λ' has different signs on both sides of the discontinuity. It means that in the given point neighbourhood, one of the two possible heating methods always causes oscillatory instability; the other method does not cause oscillatory instability.

At finite k, the formula expressing the dependence of $d\lambda/dMr$ on the system parameters turns out to be highly bulky. That is why we restrict ourselves to the asymptotic case consideration.

Within small k limit we get

$$\frac{d\lambda}{dMr} = \frac{s}{Mr_m^2} \frac{(1+\eta a)(1+\kappa a)^2}{Pa^2\kappa} \left\{ \frac{19}{25200}(1-\nu\chi a^4)+ \right.$$

$$+ (1-\chi a^2)\left[\frac{a(\eta+\nu a)}{600(1+\eta a)} + \frac{a(\kappa+\chi a)}{40(1+\kappa a)} + \right. \qquad (2.60)$$

$$\left. \left. + \frac{(1+\chi P\kappa a^3)P}{120} - \frac{P(1+\chi a^2)}{252} \right] \right\}.$$

In the region of the parameter a values close to $a_* = \chi^{-1/2}$ (see Subsection 1) formula (2.60) is simplified:

$$\frac{d\lambda}{dMr} = \frac{s}{Mr_m^2} \frac{(1+\eta a)(1+\kappa a)}{Pa^2\kappa} \frac{25200}{19} \frac{1}{1-\nu\chi^{-1}}.$$

The condition for oscillations $d\lambda/dMr > 0$ is fulfilled in case $\nu > \chi$ for heating from above ($s = -1$) and, in case $\nu < \chi$, it is fulfilled for heating from below. In other words, oscillations arise in a longwave region if heating is performed from the fluid with a large Prandtl number.

Let us consider the shortwave region ($k \to \infty$). We shall remind of the fact (see Fig. 2.19) that in this region monotonous instability arises when heating from above ($s = -1$) if $\chi < 1$ and when heating from below ($s = 1$) if $\chi > 1$ (it corresponds to the heating from the fluid with lower temperature diffusivity). The expression for $d\lambda/dMr$ within large k limits is written:

$$\frac{d\lambda}{dMr} = \frac{sk^3}{Mr_m^2} \frac{32\nu(1+\kappa)(1+\eta)(1+\kappa a)}{P^2\chi\kappa} \times$$

$$\left\{ (1+\nu\chi) + (1-\chi)\left(\frac{\eta+\nu}{1+\eta} + 4P \right) \right\}^{-1}.$$

Hence we can come to the conclusion that for $\chi < 1$ and heating from above, there is no shortwave oscillatory instability and for heating from below oscillatory instability arises if $1 < \chi < \chi_m$, where

$$\chi_m = 1 + (1+\nu\chi)\left(\frac{\eta+\nu}{1+\eta} + 4P \right)^{-1}.$$

Another criterion of oscillatory instability can be obtained if a shortwave asymptotic neutral curve is considered.

Let us introduce variable $Z = kz$ and use $\vartheta_m = kT_m$, $\lambda_i = k^2\Omega$, $Mr = k^2 M$. The system (2.53)–(2.55) for neutral disturbances ($\lambda_r = 0$) gets the form:

$$c_m(\psi_m^{IV} - 2\psi_m'' + \psi_m) = -i\Omega(\psi_m'' - \psi_m),$$

$$\frac{d_m}{P}(\theta_m'' - \theta_m) + i\psi_m A_m = -i\Omega\theta_m, \tag{2.61}$$

$$Z = k : \psi_1 = \psi_1' = \theta_1 = 0; \quad Z = -ka : \psi_2 = \psi_2' = \theta_2 = 0, \tag{2.62}$$

$$Z = 0 : \psi_1 = \psi_2 = 0, \psi_1' = \psi_2', \theta_1 = \theta_2, \kappa\theta_1' = \theta_2', \tag{2.63}$$

$$\eta\psi_1'' - \psi_2'' - iM\theta_1 = 0. \tag{2.64}$$

Since shortwave disturbances are localized close to the interface, in case $k \to \infty$ boundary conditions (2.62) can be substituted in the following way:

$$Z \to \infty : \quad \psi_1, \theta_1 \to 0; \quad Z \to -\infty : \quad \psi_2, \theta_2 \to 0. \tag{2.65}$$

The boundary problem (2.59), (2.61)–(2.63) is much like the problem of two infinite fluid layers system when mass transfer through the interface takes place, investigated for the first time in [174]. So the problem (2.59), (2.61)–(2.63) reduces to the above mentioned one if we set $\kappa = \chi = D$, where D is the diffusion coefficients ratio in the first and the second media.

The equation system solution (2.61) satisfying the boundary conditions (2.62), (2.63) is written in the form:

$$\psi_1 = e^{-Z} - e^{-r_1 Z}, \quad \psi_2 = -\frac{1-r_1}{1-r_2}\left(e^Z - e^{r_2 Z}\right),$$

$$\theta_1 = h_1 e^{-r_3 Z} - \frac{iPA_1}{1-r_3^2}e^{-Z} + \frac{iPA_1}{r_1^2 - r_3^2}e^{-r_1 Z}, \tag{2.66}$$

$$\theta_2 = h_2 e^{r_4 Z} + i\chi PA_2\frac{1-r_1}{1-r_2}\frac{1}{1-r_4^2}e^Z - i\chi PA_2\frac{1-r_1}{1-r_2}\frac{1}{r_2^2 - r_4^2}e^{r_2 Z},$$

where

$$r_1 = \sqrt{1 - i\Omega}, \quad r_2 = \sqrt{1 - i\Omega\nu}, \quad r_3 = \sqrt{1 - i\Omega P}, \quad r_4 = \sqrt{1 - i\Omega\chi P};$$

explicit coefficients h_1 and h_2 are too cumbersome to quote.

Substituting (2.66) in (2.64) we get:

$$sM = f(\Omega),$$

$$f(\Omega) = \frac{1 + \kappa a}{\kappa P} \frac{(r_4 + \kappa r_3)[\eta(1 + r_1) + (1 + r_2)]}{\chi(r_4 + 1)^{-1}(r_2 + r_4)^{-1} - (1 + r_3)^{-1}(r_1 + r_3)^{-1}}. \quad (2.67)$$

Equation (2.67) contains two equalities: equality

$$\mathrm{Im} f(\Omega) = 0 \qquad (2.68)$$

defining the oscillation frequency dependence on the system parameters, and the equation

$$sM = \mathrm{Re} f(\Omega), \qquad (2.69)$$

describing shortwave asymptotics of a neutral curve.

The equation solution (2.68) in general can only be obtained numerically. The function $\mathrm{Im} f(\Omega)$ asymptotic to $\Omega \to 0$ and $\Omega \to \infty$ being analysed analytically, one can state sufficient conditions for the oscillatory instability existence. It has the form in the region of small Ω

$$f(\Omega) = sM_0(1 - i\Omega f_1),$$

where

$$sM_0 = \frac{8(1 + \kappa a)(1 + \kappa)(1 + \eta)}{\kappa P(\chi - 1)},$$

$$f_1 = \frac{P}{2}\left(\frac{\kappa + \chi}{1 + \kappa} + 1 + \chi\right) + \frac{1}{4}\frac{\eta + \nu}{\eta + 1} + \frac{1}{4}\frac{\chi\nu - 1}{\chi - 1}. \quad (2.70)$$

The expression for f_1 coincides with the one obtained in [174] as $\kappa = \chi = D$.

In the limit $\Omega \to \infty$ we find:

$$\mathrm{Im} M = -\Omega^{3/2} f_2,$$

$$f_2 = \frac{1 + \kappa a}{\kappa\sqrt{2}}\chi^{1/2}(\chi^{1/2} + \kappa)(\nu^{1/2} + \eta)(\nu^{1/2} + \chi^{1/2}P^{1/2})(1 + P^{1/2})\times$$

$$\frac{[(\eta + 1)(\eta + \nu^{1/2})^{-1} + P^{-1/2}](\chi^{1/2} - \nu^{1/2}) + (1 - \chi^{1/2})(1 + P^{-1/2})}{(\chi^{1/2} - \nu^{1/2})^2}.$$

$$(2.71)$$

A shortwave oscillatory instability definitely exists if $sM_0 f_1$ and f_2 have opposite signs.

3. Oscillatory instability (numerical results). Let us describe numerical procedure to obtain the oscillatory neutral curve. At fixed parameters Mr and $\lambda = i\omega$ values six linear independent solutions of the equations system (2.53) satisfying the boundary conditions (2.54) have been constructed by the Runge-Kutta method. The conditions (2.55) are equivalent to some complex determinant $\Delta(Mr, \omega)$ being equal to zero. Selection of corresponding values Mr and ω has been performed by the Gauss method.

Let us present the results of neutral curves calculation carried out for transformer oil-formic acid system. Fig. 2.22 presents the critical value sMr and the frequency ω as a function of the wave number k for $a = 1.667$. Let us pay attention to the following important case: for thermocapillary convection oscillation frequency in a longwave region goes to a constant value. In the final point of the oscillatory neutral curve on the monotonous curve the oscillatory frequency ω vanishes. It is worth mentioning that in accordance with point 2 it is heating from the fluid with the large Prandtl number side (transformer oil) that causes oscillation.

Let us consider special case $\chi = 1$, $a = 1$ discussed in Subsection 1. This case is of interest because there is no monotonous neutral curve in it, and owing to this thermocapillary oscillations are the only source of instability. Fig. 2.23 presents the calculation results of the oscillatory neutral curve and the frequency dependence on the wave number for parameters $\eta = \nu = 0.5$; $\kappa = \chi = P = a = 1$ [103].

2.6. Combination of mechanisms of instability

Let us turn to the investigation of convection in a two-layer system due to the combined action of Rayleigh (volume) and thermocapillary (surface) mechanisms of instability [60]. The problem is described by the equations system (2.14), (2.15) with boundary conditions (2.16), (2.17), (2.24)–(2.28).

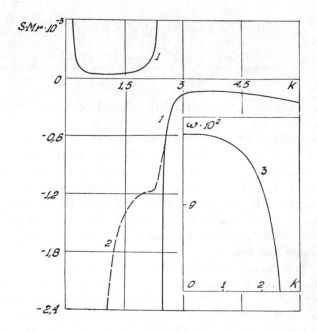

Figure 2.22. Transformer oil-formic acid system. Monotonous (1) and oscillatory (2) neutral curves; frequency ω dependence on wave number k (3).

For many systems, the "local" Rayleigh numbers $R = g\beta_m|A_m|a^4/\nu_m\chi_m$ ($m = 1, 2$) characterizing convection in each fluid, differ considerably. That is why close to the threshold thermogravitational convection arises mainly in one fluid; a weak induced flow arises in the other fluid (Section 2.2).

Thermocapillary effect acts differently on the convection in the upper and lower fluids. Let convection be realized mainly in the lower fluid (Fig. 2.24a). Above the fluid up-current the interface temperature increases causing decrease of the surface tension coefficient. Tangential stresses being the result of the mentioned above process increase the motion intensity. In this case, thermocapillary effect taking place, the critical Grashof number is expected to decrease. On the contrary, if convection is realized in the upper fluid mainly (Fig. 2.24b), thermocapillary forces prevent the fluid from moving. In this case the monotonous instability threshold increases.

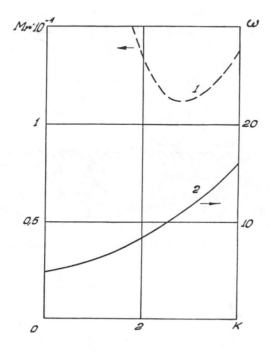

Figure 2.23. Oscillatory neutral curve (1) and frequency ω dependence on wave number k (2) for the system with parameters $\eta = \nu = 0.5$, $\kappa = \chi = P = a = 1$.

There are some fluids (for example the above transformer oil-formic acid system) for which at some wave number values convection starts in the upper fluid, and at other values the phenomenon takes place in the lower fluid. For such systems the disturbances stabilization is expected to be natural at $Mr \neq 0$. These disturbances correspond to the convection in the upper fluid and the decrease of convection threshold in the lower fluid.

Numerical calculations prove these considerations. Let us consider an air-water system first (see Table 1) for which at $Mr = 0$, $a = 1$ convection arises in the lower fluid mainly. Instability is of monotonous character, $G_* = 200$. As Mr increases, convection threshold decreases according to a linear law, vanishing at $Mr = 21$ (see Fig. 2.25).

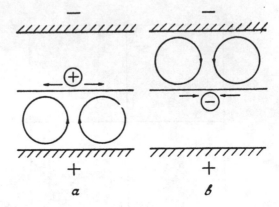

Figure 2.24. The influence of thermocapillary effect on Rayleigh convection.

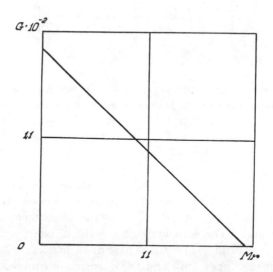

Figure 2.25. Dependence of G_* on Mr for an air-water system ($a=1$).

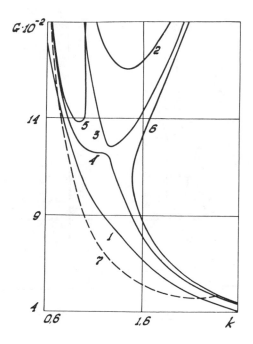

Figure 2.26. The neutral curves for a water-silicone oil No. 200 system $(a=1.6)$.

A more complicated situation is realized, for example, for a water-silicone oil DC N 200 system at $a = 1.6$. If $Mr = 0$, the threshold Grashof number for convection arising in the upper fluid is $G_1 = 270$, and in the lower fluid it is $G_2 = 2860$; both modes of instability are monotonous.

As Mr increases, the neutral curve for the disturbances in the first fluid slowly goes up, and for disturbances in the second fluid it quickly lowers. Lines 1 and 2 Fig. 2.26 correspond to monotonous neutral curves for $Mr = 175$. As Mr increases, the neutral curves become close (lines 3 and 4; $Mr = 201$). Later on, coupling and then their division into separate parts lying in the longwave and the shortwave regions (lines 5 and 6 respectively; $Mr = 210$) takes place.

The dependence of the Grashof number minimized with respect to k on Mr number for the lower neutral curve (line 1) and for the upper

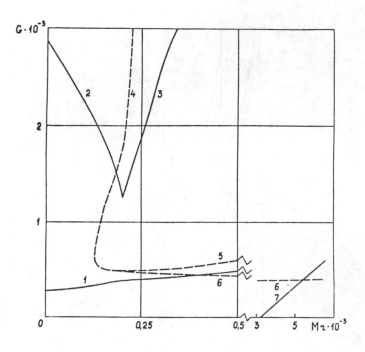

Figure 2.27. Dependence of G_* on Mr for a water-silicone oil No. 200 system (a=1.6).

neutral curve before and after coupling (lines 2 and 3 respectively) is shown in Fig. 2.27.

Let us turn now to the discussion of the oscillatory instability. At $Mr > 125$ on the neutral curve for the disturbances in the upper fluid an oscillatory zone appears. It is kept after monotonous neutral curves couple, connecting longwave and shortwave fragments of the monotonous neutral curve. For $Mr = 210$ the oscillatory neutral curve is shown in Fig. 2.26 by dashed line 7; for other Mr values the oscillatory neutral curves are not shown because they coincide with the shown curve for this graph scale.

In Fig. 2.27 the dependences on Mr for the Grashof numbers corresponding to the left and right final points of oscillatory neutral curve (lines 4 and 5) and the minimum Grashof number for oscillatory instability (line 6) are shown. In the range $Mr > 400$ oscillatory instability becomes the most dangerous one. At $Mr > 3200$, $G = 0$ monotonous

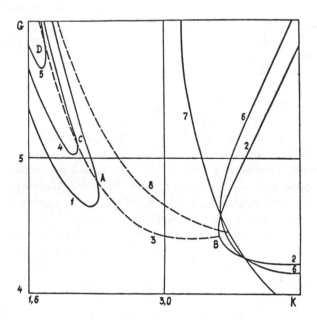

Figure 2.28. The neutral curves for a transformer oil-formic acid system (a=0.667).

thermocapillary instability of the reference state takes place. As G increases, thermocapillary instability stabilizes. The steady state turns out to be stable in a finite interval of G values, corresponding to the region between lines 7 and 6 of Fig. 2.27. At $Mr > 5300$ the steady state becomes unstable at any G values.

Let us consider now convection generation in a transformer oil-formic acid system. As mentioned in Section 2.2, for any interval of values a the monotonously increasing disturbances with small wave numbers are localized mainly in the upper fluid, and the shortwave disturbances are localized in the lower one. Depending on the layers' thickness ratio, the minimum of the neutral curve may be realized both at shortwave ($a = 0.667$) and longwave ($a = 0.54$) disturbances. In an intermediate wave number region the oscillatory instability takes place (being, however, not the most dangerous one).

We shall start with the case $a = 0.667$ for which as $Mr = 0$ convection arises in the lower fluid first. The corresponding neutral curves are shown in Fig. 2.28; the monotonous neutral curves are shown in solid lines and oscillatory branches of neutral curve are shown in dashed lines. Lines 1 and 2 and part of line 3 between points A and B correspond to the value $Mr = 0$. Even for not large Mr values the left fragment of monotonous neutral curve and the left final point of oscillatory branch 3 quickly move to the longwave region. Lines 4 and 5 show parts of monotonous neutral curves at $Mr = 0.5$; 1; points C and D are the final points of oscillatory branch 3 at the same values of Mr. The deviation of the oscillatory branch from the case $Mr = 0$ is not visible on the graph scale. The right fragment of the neutral curve when Mr increases changes considerably slowlier; line 6 corresponds to $Mr = 5$, lines 7 and 8 correspond to $Mr = 100$. The threshold Grashof number G_* corresponding to the disturbances in the lower fluid, which can be obtained by a minimization of the Grashof number according to the wave number, monotonously lowers when Mr increases.

Let us describe now the case $a = 0.54$ for which at $Mr = 0$ the most dangerous are the longwave disturbances in the upper fluid (see Fig. 2.29, lines 1–3). As in the previous case, the left part of the monotonous neutral curve when Mr increases considerably stabilizes, causing an expansion of the oscillations existence region. At $Mr = 10$, a minimum appears on the oscillatory branch of neutral curve, and at $Mr > 11.5$ oscillatory disturbances become the most dangerous ones. The neutral curves at $Mr = 20$ are shown in Fig. 2.29 (lines 4–6). As Mr grows, the oscillatory convection threshold increases and the monotonous instability threshold corresponding to the convection development in the lower fluid decreases (see lines 7 and 8 in Fig. 2.29; $Mr = 200$). As Mr increases further, the monotonous disturbances again become the most dangerous ones. The dependence of oscillations frequency on the wave number is shown in Fig. 2.30.

The dependence of the Grashof number minimized according to k on Mr number corresponding to monotonously increasing longwave (line 1), shortwave (line 2) and oscillatory (line 3) disturbances are shown in Fig. 2.31. Lines 4 and 5 correspond to the left and right final points of the oscillatory branch.

Thus, the influence of thermocapillary effect on the convection appearance in a two-layer system depends considerably on the disturbances form, causing instability. If critical disturbances are localized in the lower fluid, then the effect leads to the decrease of the monotonous instability threshold. If the disturbances are localized in the upper fluid, the inclusion of thermocapillary effect leads to the fact that the monotonous instability stabilizes and the oscillatory mode becomes the most dangerous one.

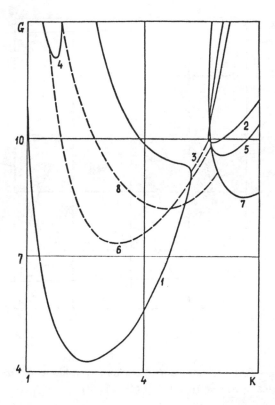

Figure 2.29. The neutral curves for a transformer oil-formic acid system $(a=0.54)$.

In conclusion we shall mention the articles [80, 81] devoted to theoretical and experimental investigation of convection arising at combined action of both mechanisms of instability for benzene-water and water-carbon-tetrachloride carbon systems. Calculations were carried out under assumption of a nondeformable interface; only the monotonous mode of instability, which the existence was confirmed experimentally, has been investigated.

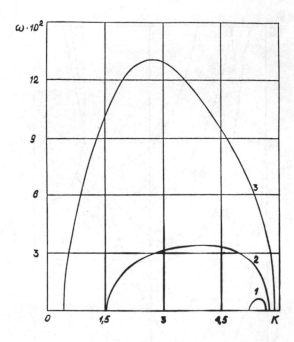

Figure 2.30. Dependence of ω on k for a transformer oil-formic acid system $(a=0.54)$; $Mr=0$ (line 1), 20 (2), 200 (3).

2.7.   The effects of interface deformation

In previous paragraphs, we considered the problem of a two-layer system stability under assumption of a flat interface. The conditions at which interface deformation may be neglected are presented in Section 1.2.

Let us remind that in order to calculate the thermogravitational convection threshold $(G \neq 0, Mr = 0)$, in the framework of Boussinesq approach, it is not necessary to take into account the interface deformation except if the media have close densities $(|\rho - 1| \ll 1)$. For thermocapillary convection $(Mr \neq 0)$ in the longwave region the interface deviation is the dominating factor and the problem must be solved in full.

1. Thermogravitational convection in media with close densities. Let us consider thermogravitational convection in media with close densities described by the boundary value problem (2.14)–(2.23) (in the boundary condition (2.19) we assume $Mr = 0$).

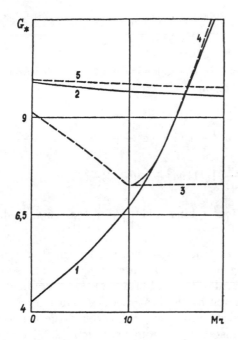

Figure 2.31. Dependence of G_* on Mr for a transformer oil-formic acid system (a=0.54).

Let us first underline that the interface deformability ($h \neq 0$) causes a new mode of disturbance leading to a longwave system instability. It is not difficult to prove that at $k = 0$ three types of normal disturbances are realized. Two of them make up the account set of decreasing temperature disturbances and decreasing stream function disturbances keeping the interface flat. Moreover, a special neutral mode exists ($\lambda = 0$) :

$$h^{(0)} = 1, \quad T_1^{(0)} = \frac{s(1 - \kappa)}{(1 + \kappa a)^2}(1 - z),$$

$$T_2^{(0)} = -\frac{s\kappa(1 - \kappa)}{(1 + \kappa a)^2}(z + a),$$

corresponding to the homogeneous displacement of the interface. For this mode, using expansions in the wave number:

$$h = 1, \quad T_m = T_m^{(0)} + k^2 T_m^{(2)} + \cdots, \psi_m = k\psi_m^{(1)} + k^3\psi_m^{(2)} \cdots,$$
$$\lambda = k^2\lambda^{(2)} + k^4\lambda^{(4)} + \cdots,$$

it is possible to obtain the following expression defining the decrement in the longwave region

$$\lambda^{(2)} = \left\{ \frac{1}{3}\eta a^3 Ga(\rho^{-1} - 1)(1 + \eta a) + \frac{1}{3}sG\eta a^3 \frac{(1 - \rho^{-1}\beta^{-1})(1 + \eta a)}{1 + \kappa a} - \right.$$
$$\frac{sG(1 - \kappa)a^2}{120(1 + \kappa a)^2}\left[\eta(11\eta a^2 + 14a + 3) - \right.$$
$$\left. \left. \kappa\nu a^3\beta^{-1}(3\eta a^2 + 14\eta a + 11) \right] \right\}(1 + 4\eta a + 6\eta a^2 + 4\eta a^3 + \eta^2 a^4)^{-1}.$$

$$(2.72)$$

The first term in braces describes the known Rayleigh-Taylor instability developping in isothermal case at $\rho > 1$. The second term is a correction connected with equilibrium pressure gradients change close to the interface given by the heat media expansion.

Finally, the third term takes into account the convection effect on Rayleigh-Taylor instability. From expession (2.72) one can see, that depending on media parameters and types of heating the corrections may have both a stabilizing and a destabilizing character.

For finite k the calculation of the disturbances spectrum requires, generally speaking, a numerical solution of the system (2.14)–(2.23) and has so far been done only for some particular cases. In articles [139, 141] a model system with free external boundaries stability where the parameters $\rho, \kappa, \chi, \nu, \beta$ are close to unity and the variable W is close to zero, is investigated. Simplification of the problem is connected with the fact that, at $\rho = \kappa = \beta = 1$ and $W = 0$, the interface deformation does not lead to mechanical equilibrium disturbance at any k at all, so that a neutral mode of disturbances exists: $h = 1, \psi_m = 0, T_m = 0$ $(m = 1, 2)$. At small deviations of the system parameters from unity, perturbation theory can be applied for calculation of this decrement. In [139] the case $G < G_c$ (G_c is the threshold Grashof number in the absence of boundary deformations) is under consideration; in this region the instability is of a Rayleigh–Taylor nature and has a monotonous character. It is stated that corrections to the decrement values connected with convection may have both a stabilizing and a nonstabilizing character. In [141] the case $G = G_c$ is studied, where together with modes corresponding to the interface deformation, the convective mode corresponding to the undeformable interface is also neutrally stable, the system parameters being equal to unity. It is shown that for parameters different from unity, an oscillatory instability is possible as a result of both modes interaction. Analytical results are in agreement with numerical calculations performed in [140].

In [136] for the chosen model system one observes, the existence of oscillatory neutral curve, connected with the interface distortion lying in the region of longer waves than the ones corresponding to monotonous neutral curves. For any fixed wave number the dependence of the critical Rayleigh number on the ratio of media densities (close to $\rho = 1$) is investigated.

2. Thermocapillary convection. Monotonous mode of instability. Let us consider now the case of thermocapillary convection, described by the following boundary value problem:

$$c_m D^2 \psi_m = -\lambda D \psi_m,$$
$$\frac{d_m}{P} D T_m + ik A_m \psi_m = -\lambda T_m \quad (m = 1, 2), \tag{2.73}$$

$z = 1: \ \psi_1 = \psi'_1 = T_1 = 0; \quad z = -a: \ \psi_2 = \psi'_2 = T_2 = 0$

$z = 0: \ \psi'''_1 - \eta^{-1}\psi'''_2 + \left[\lambda(1 - \rho^{-1}) - 3k^2(1 - \eta^{-1})\right]\psi'_1 +$

$\qquad ik\left[Ga(\rho^{-1} - 1) + Wk^2\right]h = 0$

$\qquad \eta(\psi''_1 + k^2\psi_1) - (\psi''_2 + k^2\psi_2) - ikMr\left(T_1 - \dfrac{s}{1 + \kappa a}h\right) = 0$

$\qquad \psi'_1 = \psi'_2$

$\qquad \psi_1 = \psi_2 = -i\dfrac{\lambda}{k}h$

$\qquad T_1 - T_2 = \dfrac{s(1 - \kappa)}{1 + \kappa a}h$

$\qquad \kappa T'_1 - T'_2 = 0.$

In equations (2.73) we assume $G \to 0$ at finite Ga value, corresponding the limit $\delta_\beta \to 0$.

The threshold of monotonous instability ($\lambda = 0$), like in the case of an undeformable interface can be calculated analytically [169]. The solution of the problem is still described by (2.56); the only difference is that coefficients e_1 and e_2 get the following increase Δe_1 and Δe_2:

$$\Delta e_1 = -\frac{s(1 - \kappa)}{(1 + \kappa a)(t_1 + \kappa t_2)}h,$$

$$\Delta e_2 = -\frac{s\kappa(1-\kappa)}{(1+\kappa a)(t_1 + \kappa t_2)} h,$$

where the interface deformation h is determined by the exression

$$h = -2ik^2 \left[1 - \frac{(S_1^2 - k^2)a^2}{\eta(S_2^2 - k^2 a^2)} \right] (Ga\delta + Wk^2)^{-1},$$

$$\delta = \rho^{-1} - 1.$$

The neutral curve can be calculated by the formula

$$Mr(k) = \frac{8sk^2(1+\kappa a)(\kappa D_1 + D_2)(\eta B_1 + B_2)}{\kappa[P(\chi E_2 - E_1) - 8k^5(D_1 + D_2)(F_1 - F_2\eta^{-1})(Ga\delta + Wk^2)^{-1}]};$$
$$(2.74)$$

here $F_1 = (S_1^2 - k^2)^{-1}$, $F_2 = a^2(S_2^2 - k^2 a^2)^{-1}$; the other notations have been used in the formula (2.57).

At $k \to \infty$ the expression (2.74) turns into (2.57), i.e. within short-wave limit the deformation of the interface is not essential. On the contrary, for longwave disturbances the distortion of the interface is of dominating significance. At $k = 0$ the convection arising threshold is described by the expression [169]:

$$Mr(0) = \frac{2s\eta Ga\delta(1+\eta a)(1+\kappa a)^2 a}{3\kappa(1+a)(1-\eta a^2)}. \qquad (2.75)$$

This leads to the fact that a longwave convection arises when heating from below ($s > 0$) if $a < \eta^{-1/2}$, and when heating from above if $a > \eta^{-1/2}$. Let us remind of the fact that when neglecting the interface distortion the longwave asymptotics of the neutral curve is described by (2.58), and the way of heating, for which the thermocapillary convection arises is determined by the variables a and $\chi^{-1/2}$. In the case when the variables $\eta a^2 - 1$ and $1 - \chi a^2$ have different signs the formulas (2.58) and (2.75) give opposite predictions concerning the convection behaviour.

Let us discuss the validity region of the formula (2.58). The analysis of the general expression (2.74) shows that the formula (2.58) may be applied in any region of wave numbers $k_0 \ll k \ll 1$ if the parameters $Ga\delta$ or W are large. In case $Ga\delta \gg 1$ we obtain the estimate $k_0 \sim (Ga\delta)^{-1/2}$; at $Ga\delta \sim 1$, $W \gg 1$ we have $k_0 \sim W^{-1/4}$. In the region of wave numbers $k \leq k_0$ the formula (2.74) may be approximately written in the form [169]:

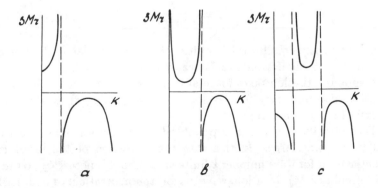

Figure 2.32. The change of the neutral curve form with the increase in parameter a as $\eta < \chi < 1$: a) $a < 1/\sqrt{\chi}$; b) $1/\sqrt{\chi} < a < 1/\sqrt{\eta}$; c) $a > 1/\sqrt{\eta}$.

$$Mr(k) = -\frac{80s(1 + \eta a)(1 + \kappa a)^2 \eta a}{\kappa\{\eta a^3 P(1 - \chi a^2)k^2 + 120(1 + a)(\eta a^2 - 1)[Ga\delta + Wk^2]^{-1}\}}.$$
$$(2.76)$$

The expression (2.76) "sews together" asymptotic formulas (2.58) and (2.75). It is seen that in the case when the values $\eta a^2 - 1$ and $1 - \chi a^2$ turn out to be of opposite sign, the neutral curve $sMr(k)$ has a discontinuity in the longwave region at some wave number value $k^2 = k_*^2$, defined by the annulation of the denominator in the expression (2.76). If the values $\eta a^2 - 1$ and $1 - \chi a^2$ are of the same sign, the neutral curve in the longwave region is continuous.

As an example let us describe the change of the neutral curve form for the system with parameters $\eta < \chi < 1$. At $a < 1/\sqrt{\chi}$ the neutral curve in the longwave region has a discontinuity; in the case $1/\sqrt{\chi} < a < 1/\sqrt{\eta}$ the discontinuity is realized as $k \sim 1$; if $a > 1/\sqrt{\eta}$, two discontinuities exist: in the longwave and the shortwave regions (see Fig. 2.32).

The above reason correspond to the case $a \sim 1$. If $a \ll 1$ then the conditions $Ga\delta \gg 1$ or $W \gg 1$ can turn out to be not sufficient for Equ. (2.58) applicability. For example, if $Ga\delta \sim 1$, $W \gg 1$ then the position of the longwave neutral curve discontinuity point according to Equ. (2.76) is described by the formula

$$k_*^4 \simeq \frac{120}{W a^3 \eta P},$$

so at sufficiently small a the condition $k_* \ll 1$ is violated. In this case even at $k \sim 1$, the formula (2.74) cannot be simplified.

A similar situation takes place at $a \gg 1$ when the condition $k_* a \ll 1$ which is also necessary for developping asymptotic expressions in the formula (2.74) is not fulfilled.

Theoretically a situation is possible for which both parameters $Ga\delta$ and W are not large. In this case the distortion of the interface is essential even for wave number k finite values, so it is necessary to use the full formula (2.74). In a longwave region the neutral curve is described by the expression

$$Mr(k) = Mr(0)(1 + k^2 N),$$

where

$$N = \frac{(\kappa - 1)a(1 - a)}{3(1 + \kappa a)} + \frac{(\eta + a)a}{15(1 + \eta a)} +$$
$$\frac{2(1 - \eta)a^2}{15(1 - \eta a^2)} + \frac{W}{Ga\delta} + \frac{P(1 - \chi a^2)\eta a^3 Ga\delta}{120(1 + a)(1 - \eta a^2)}. \quad (2.77)$$

Depending on the values of system parameters at $k = 0$, a minimum ($N > 0$) or a maximum ($N < 0$) of a neutral curve can be realized. It is necessary to remember that for real fluids the relative change of surface tension stipulated by temperature heterogeneity is usually not large: $\delta_\alpha = Mr/W\eta \leq 1$. That is why, only that part of the neutral curve which satisfies this inequality, has physical sense.

In conclusion we shall remind of the fact that in the article [98] a modification of the condition for heat fluxes on the interface, taking into account the energy expenditure on the deformation of the interface is suggested (see Section 1.2). Stability analysis taking into consideration this effect is performed in [9]. It is stated, that the energy expenditure effect on the interface deformation is determined by the parameter $\Lambda = \alpha^2 T_0 / \eta_2 \kappa_2$, where T_0 is the absolute interface temperature. The analytical formula defining the marginal stability threshold with respect to the monotonous disturbances, transforming in (2.74) as $\Lambda \to 0$, is obtained. The estimates [9, 135] show, however, that under real conditions the parameter Λ is usually small.

3. Thermocapillary convection. Oscillatory mode of insta-bility. The monotonous mode of instability is naturally, not the only one. As shown in Section 2.5, even if interface deformation is absent, there are situations when an oscillatory mode of instability arises. Apart from this, we can expect the appearance of specific oscillations connected with a wave arising on the interface. In the case of a one-layer system, similar oscillations are considered in [157].

As the influence of the interface deformation is assumed to be the strongest one for the case of fluids with nearly equal densities, let us first consider the model system with $\rho = 0.999$. Other parameters of the system are chosen as follows: $\chi = a = W = 1$ and $\nu = 0.5$. As stated in Section 2.5, in the absence of interface deformation the only possible mode of instability is the oscillatory mode, realized for heating from below $(s = 1)$.

At finite values of Ga for this system heated from below, the monotonous mode of instability also appears; the critical value of Mr corresponds to $k \to 0$ and, according to (2.75), is $Mr(0) \simeq 2 \cdot 10^{-3} Ga$. Neutral curves at finite Ga values are shown in Fig. 2.33. The de-pendence of the oscillation frequency on the wave number is shown in Fig. 2.34. One can see that the oscillatory instability disappears in the longwave region and is kept only for values of the wave number k larger than $k_*(Ga)$; at $k = k_*$ the oscillatory frequency vanishes. With de-crease of the parameter Ga, the monotonous neutral curve falls more rapidly than the oscillatory one so that the longwave monotonous dis-turbances, essentially connected with the interface deformation, become more dangerous (at $Ga \leq 5 \cdot 10^6$).

The deformation of the interface can lead to the onset of oscillatory instability even in the case when without deformation the steady state is absolutely stable [112]. Let us consider, for example, the system heated from above, described earlier. For the system with an interface defor-mation, a new type of oscillatory instability is discovered. The neutral curve looks like a "sack" (Fig. 2.35 a), i.e. two values of sMr corre-spond to any wave number in some region $k < k_1(Ga)$. The dependence of the oscillatory frequency on the wave number k for both branches of the neutral curve is shown in Fig. 2.35 b. As $k \to 0$, the upper and the lower branches of the neutral curve, marked with "+" and " − ", have asymptotic limits $Mr \to C_\pm k^{-2}$ and $\omega \to \omega_\pm$; moreover, the values $C_\pm, \omega_\pm$ do not depend on the parameter Ga. The threshold value of Mr tends to infinity as $Ga \to \infty$; this result proves that the instability is con-nected with the interface deformation. The width of the interval in wave number k_1 unstable region decreases monotonously with increasing Ga.

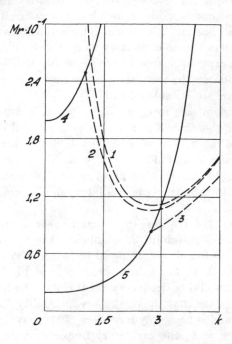

Figure 2.33. Oscillatory $(Ga=10^9$ — curve 1; 10^7 — 2; 10^6 — 3) and monotonous $(10^7$ — 4; 10^6 — 5) neutral curves for the model system.

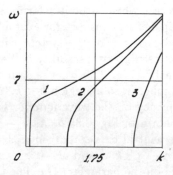

Figure 2.34. Dependence of the frequency on the wave number $(Ga=10^9$ — curve 1; 10^7 — 2; 10^6 — 3).

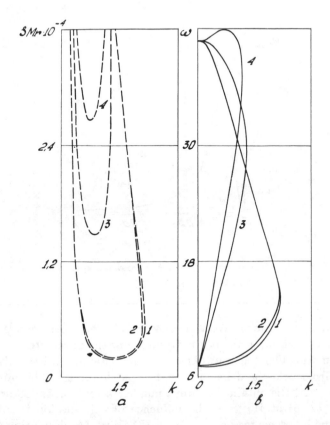

Figure 2.35. Dependence of the parameter sMr (a) and of the frequency ω (b) on the wave number ($Ga=0$ — curve 1; 10^4 — 2; 10^6 — 3; $2 \cdot 10^6$ — 4).

Thus, in the system considered, an oscillatory instability is realized for both types of heating.

Let us turn to the thermocapillary convection in the real air–water system (Table 1). We shall consider the case $a = 1$. For this system on the Earth, the parameters W and Ga are not independent and are connected with each other by the relation $Ga = (W/\gamma)^3$, where the constant $\gamma = (\Sigma/\rho_1)(g\nu_1^4)^{-1/3}$ depends upon the physical properties of the media only. However, we shall consider Ga and W as independent parameters, with g, the gravity acceleration, supposed to change under

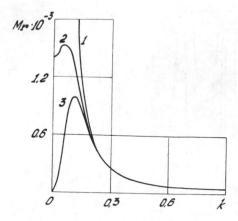

Figure 2.36. The neutral curves for an air-water system heated from below $(Ga=10^6$ — curve 1; 10 — 2; 0 — 3).

microgravity conditions. We choose $W = 10^6$ (that corresponds to the thickness of each layer about 3 mm) and vary the parameter Ga.

Deformation being absent $(Ga \to \infty)$, an instability arises only when heating from below and has a monotonous character. The minimum value $Mr_c \simeq 21$ is obtained at wave number $k_c \simeq 2.0$. As follows from equation (2.76) at $W \gg 1$, the influence of deformation is sufficient only in the longwave region (see Fig. 2.36). With Ga decreasing (in the region $Ga < Ga_1 \simeq 13.5$), the second minimum corresponding to $k = 0$ arises on the neutral curve. The longwave disturbances are the most dangerous if $Ga < Ga_2 \simeq 0.15$.

In the air-water system heated from above the steady state for the case of a flat interface at $a = 1$ is absolutely stable. The deformation of the interface leads to a new type of the oscillatory instability which differs from the one that was discussed earlier. Formally, neutral curves can be calculated for any value of Ga. It is necessary to remember, however, that the temperature change of the surface tension $\alpha\theta$ must be larger than its mean value $\Sigma : \alpha\theta \leq \Sigma$; this leads to relation $Mr \leq W\eta$. According to calculations this relation is satisfied for the system under consideration only if $Ga \leq 20$, i.e. under microgravity conditions.

Oscillatory neutral curves and the dependeces of frequency on the wave number are shown in Fig. 2.37 and 2.38 respectively. Critical wave numbers lie in the longwave region. It is necessary to emphasize that

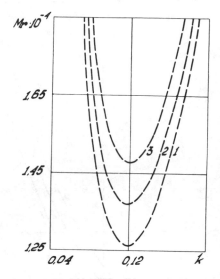

Figure 2.37. The neutral curves for an air-water system heated from above $(Ga=0 - 1; 3 - 2; 6 - 3)$.

in this case the neutral curves are not in the form of a "sack", i.e., at $Mr \to \infty$ the width of the instability region does not decrease.

Let us discuss the reasons for the different shapes of the oscillatory neutral curves considered earlier. For that we shall calculate the sign of the longwave thermocapillary oscillations decrement in the limit $k \to 0$, $M \equiv Mrk^2 \to \infty$. It is easy to see that the frequency of oscillations in this limit obeys the relation $\omega \sim M$. Because of this the motion has the character of boundary layers near the interface and solid surfaces. Introducing notation

$$T_m = k\theta_m, h = kH, \psi_m = M\varphi_m,$$

$$\lambda = \Lambda M, z = M^{1/2}z,$$

we write the system (2.73) in the lowest order in k:

$$\Lambda\varphi_m'' = -c_m\varphi_m^{IV}, -\Lambda\theta_m - iA_m\varphi_m = \frac{d_m}{P}\theta_m'',$$

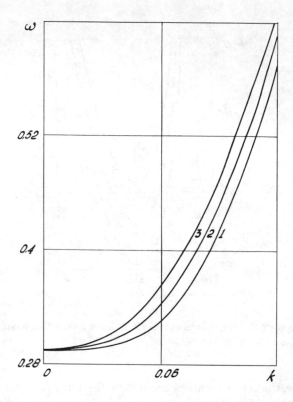

Figure 2.38. Dependence of the frequency on the wave number (the notation is the same as in Figure 2.37).

$$Z = M^{1/2} : \varphi_1 = \varphi_1' = \theta_1 = 0; z = -M^{1/2} : \varphi_2 = \varphi_2' = \theta_2 = 0,$$
$$Z = 0 : \varphi_1''' - \eta^{-1}\varphi_2''' + (1 - \rho^{-1})\varphi_1' = 0,$$
$$\eta\varphi_1'' - \varphi_2'' - i\theta_1 + i\frac{s}{1 + \kappa a}H = 0,$$
$$\varphi_1' = \varphi_2', \varphi_1 = \varphi_2 = -i\Lambda H,$$
$$\theta_1 - \theta_2 = \frac{s(1 - \kappa)}{1 + \kappa a}H, \quad \kappa\theta_1' - \theta_2' = 0.$$

The solution can be obtained analytically as an expansion in $M^{-1/2}$. The result for the oscillation frequency is the expression

$$\Omega^2 = (Im\Lambda)^2 = \frac{s\kappa}{1 + \kappa a}\left(\frac{\sqrt{P}}{1 + \sqrt{P}} - \frac{\sqrt{\chi P}}{\sqrt{\nu} + \sqrt{\chi P}}\right)\frac{1}{\sqrt{P}(\eta + \sqrt{\nu})(\kappa + \sqrt{\chi})}.$$

Oscillations appear in cases of heating such that the expression on the right-hand side is positive. Thus for both studied systems the oscillatory instability is possible when heating from above. For Re we find

$$
Re\Lambda = -M^{-1/2}(2\Omega)^{-3/2} \left[\frac{s\kappa}{1+\kappa a} \frac{1}{\sqrt{\nu}+\sqrt{\chi P}} - \Omega^2 \sqrt{P}(\kappa + \sqrt{\chi}) \right] \times
$$
$$
\times \left[\frac{1+\rho a}{\rho a - \nu^{-1/2}} \frac{1-a\sqrt{\nu}}{a} - \sqrt{\nu}\frac{1+a}{a} \right] \times
$$
$$
\times \left[\sqrt{P}(\sqrt{\chi}+\kappa)(\sqrt{\nu}+\eta)\frac{1+\rho a}{\rho a - \nu^{-1/2}} \right]^{-1}.
$$

For the model system this expression gives $Re\Lambda > 0$ so that the oscillatory instability is absent at high enough values of $M = Mrk^2$. Because of this the instability region is restricted from above by the condition $M < C_+$ or $Mr < C_+ k^{-2}$. By contrast, it follows that $Re\Lambda < 0$ for the air-water system, so that the oscillatory disturbances increase at any large value $M = Mrk^2$ (neutral curve at $Mr \to \infty$ does not close).

4. Mixed convection. The investigation of mixed convection ($G \neq 0$, $Mr \neq 0$) with the interface deformation taken into account for systems with close densities was carried out in a series of papers [42, 193]. Let us underline (see point 2) that the interface deformation is not necessary to be taken into account at $k \sim 1$, as a rule, if the layers' thicknesses are of the same order. In the longwave region to the corrections of the stability criteria (2.75) connected with the parameter G are small. So, in [193] the following formula was obtained

$$
Mr(0) = \frac{2}{3}Ga(\rho^{-1} - 1)\frac{\eta a(1+\kappa a)^2(1+\eta a)}{\kappa(1+a)(1-\eta a^2)} +
$$
$$
\frac{G\nu(1-\kappa)}{60\beta}\left[\frac{11a^3 + \beta\rho\kappa^{-1}}{1+a} + \frac{14(\eta a^4 - \beta\rho\kappa^{-1})}{1-\eta a^2} \right].
$$

Taking into account that $G = \delta_\beta \cdot Ga$, one can prove that for large values of $|\rho - 1|$ the second term in this formula has the relative smallness order of δ_β, so it goes beyond the Boussinesq approach.

If, however, $a \ll 1$ then, as shown above, at $k \sim 1$ the problem (2.14)–(2.23) must be solved in full state. At $a \gg 1$ the full problem must be solved in the region of parameters values $ka \sim 1$.

In [42, 193] the stability was studied only with respect to the monotonous disturbances. Discontinuities in the obtained neutral curves makes it possible to assume that in systems under consideration an oscillatory mode of instability is also possible.

5. Convection with jump of media parameters. Here a problem is considered which is slightly different from the ones presented in the other sections of this chapter. Let the physical parameters of fluid media have a jump at any temperature T_* stipulated, for example, by a phase transition. We shall call with index 1 the values of parameters and physical variables reffering to the region $T < T_*$, and by index 2 — to the region $T > T_*$, and we shall use the same notation as in Section 1.2. Thermocapillary convection in the regions 1 and 2 is described by the same equations (1.11) and (1.12) as for immiscible fluids. The conditions on the interface, however, will be essentially different: the interface is the isotherm $T = T_*$, the fluid flow going across it.

Let the fluid, filling up the layers $-a < z < 1$ be heated from below; at the steady state $T = T_*$ at $z = 0$. The small disturbances submit to equations (2.14), (2.15). Conditions (2.16), (2.17) are still valid for solid boundaries, or else take the form

$$z = 1 : \psi_1 = \psi_1'' = T_1 = 0; z = -a : \psi_2 = \psi_2'' = T_2 = 0$$

for the case of free boundaries. Let us discuss now the conditions of parameters jump $z = h$ on the disturbed surface. As this surface coincides with the isotherm $T = T_*$, its equation can be obtained from the condition

$$z = 0 : \frac{dT_1^0}{dz}h + T_1 = 0,$$

i.e.

$$h = s(1 + \kappa a)T_1(0).$$

As the fluid flow goes over the interface, condition (2.21) is not fulfilled; instead of it the fluid flow balance condition of the form

$$z = 0 : \rho(-ik\psi_1 + \lambda h) = -ik\psi_2 + \lambda h$$

is used. If the interface is the phase transition front of the first kind then heat balance equation being written, it is necessary to take into account the latent heat of transition

$$z = 0 : \kappa T_1' - T_2' = q(-ik\psi_1 + \lambda h),$$

where q is the dimensionless parameter, characterizing heat release or heat absorption. The boundary conditions (2.18)–(2.20), (2.22) keep the former form.

The stated boundary value problem was solved for a number of particular cases, mainly in geophysical applications. In [14, 15, 40, 177] the viscous case ($\eta = \nu \neq 1, \kappa = \chi = \beta = 1$) was analysed. The most detailed investigation of the problem is carried out for the layer with free external boundaries in [14, 15] where the opportunity of two different convective flow regimes exists. In the first regime the motion is concentrated in the lower-viscosity layer, and the higher-viscosity layer has solid bulk properties. In the second regime the motion is of through out character, and convective circulation envelopes both layers. In [160] heat release on the phase transition front and density at homogeneous viscosity was taken into account.

2.8. Convective stability in enclosed vessels

Convective stability investigation of a two fluid system filling up an enclosed vessel requires solving a more complicated mathematical problem than in the case of a horizontal layer. As a matter of fact, for horizontal layers it is possible to use the substitution (2.13) and get an ordinary differential equations system (2.14), (2.15) with the boundary conditions (2.16)–(2.23). Definition of convective stability threshold in a two-layer system in enclosed vessel may be carried out only on the basis of boundary value problem solution in partial derivatives. Up to now this problem is solved for some vessels types only[3].

1. Convection in a horizontal cylinder of rectangular section. Let the horizontal cylinder shown in Fig. 2.39 be filled up with two immiscible fluids, the interface for the steady state is described by equation $z = 0$. The horizontal cylinder boundaries are assumed to be solid and isothermal:

$$z = a_1 : v_1 = 0, T_1 = 0; z = -a_2 : v_2 = 0, T_2 = \theta. \qquad (2.78)$$

On solid vertical cylinder boundaries the temperature distribution providing mechanical reference state in the system is kept:

---

[3] Some inequalities, corresponding to the problem of thermogravitational convection arising in a two fluids system, filling up the vertical cylinder of arbitrary section were obtained in [101].

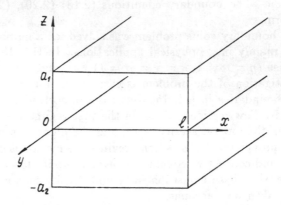

Figure 2.39. A horizontal cylinder of rectangular section.

$$x = 0, \quad x = l:$$

$$v_1 = 0, T_1 = \theta \frac{1 - za_1^{-1}}{1 + a_2\kappa_1(a_1\kappa_2)^{-1}}, z > 0; \qquad (2.79)$$

$$v_2 = 0, T_2 = \theta \frac{1 - z\kappa_1(a_1\kappa_2)^{-1}}{1 + a_2\kappa_1(a_1\kappa_2)^{-1}}, z < 0.$$

The system of convection equations (1.2) with boundary conditions (2.78), (2.79) on the vessel boundaries and (1.5)–(1.10) on the interface posseses a solution corresponding to the steady state described by the formulas (2.79) in the whole region. Let us turn to the stability investigation. We shall use the same notation and the same transition to dimensionless variables as in Sections 1.2 and 2.1. Then the dimensionless equations describing small disturbances have the form (2.2)–(2.4). On horizontal cylinder boundaries and on the interface the boundary conditions (2.5)–(2.12) hold. In contrast to the case of horizontal layers system, it is also necessary to impose boundary conditions on the vertical cylinder walls:

$$x = 0, x = L : v_m = 0, T_m = 0, \quad (m = 1, 2), \qquad (2.80)$$

where $L = l/a_1$.

As far as the linear boundary value problem (2.2)–(2.12), (2.80) is homogeneous in y and t, its solutions may be presented as a superposition of normal disturbances

$$(\tilde{v}_m(x, z), \tilde{p}_m(x, z), \tilde{T}_m(x, z), \tilde{h}(x)) \exp(iky - \lambda t)$$

(further on the sign "tilde" is omitted).

We shall restrict ourselves with the thermogravitational convection case. As mentioned in Section 1.2 in the Boussinesq approach framework, it is necessary to put $h = 0$. From the investigation results of homogeneous fluid stability in a horizontal cylinder [48] Section 20 it follows that the plane disturbances ($k = 0$) for which $v_{m,y}(x, z) = 0$, play an important role. It allows us to introduce the stream function

$$v_{m,x} = \frac{\partial \psi_m}{\partial z}, \quad v_{m,z} = -\frac{\partial \psi_m}{\partial x}.$$

Later on only the disturbances of this character will be considered. Assuming $\lambda = 0$, we get the following boundary value problem for calculation of the stability boundary with respect to monotonous disturbances

$$c_m \Delta_2^2 \psi_m - Gb_m \frac{\partial T_m}{\partial x} = 0,$$
$$\frac{d_m}{P} \Delta_2 T_m + A_m \frac{\partial \psi_m}{\partial x} = 0 \quad (m = 1, 2), \tag{2.81}$$

$$z = 1: \quad \psi_1 = \frac{\partial \psi_1}{\partial z} = T_1 = 0,$$

$$z = -a: \quad \psi_2 = \frac{\partial \psi_2}{\partial z} = T_2 = 0, \tag{2.82}$$

$$z = 0: \quad \eta \frac{\partial^2 \psi_1}{\partial z^2} = \frac{\partial^2 \psi_2}{\partial z^2}, \frac{\partial \psi_1}{\partial z} = \frac{\partial \psi_2}{\partial z}, \psi_1 = \psi_2 = 0,$$

$$T_1 = T_2, \kappa \frac{\partial T_1}{\partial z} = \frac{\partial T_2}{\partial z};$$

$$x = 0, L: \quad \psi_1 = \frac{\partial \psi_1}{\partial x} = T_1 = 0 \quad (z > 0),$$

$$\psi_2 = \frac{\partial \psi_2}{\partial x} = T_2 = 0 \quad (z < 0).$$

The solution of the boundary value problem may be carried out by numerical methods. In [22] the calculation methods based on the application of the finite difference method are suggested; in [23] for the solution of boundary value problem the finite element method was used.

Let us describe the main stages of the problem solution by the finite difference method. Changing the variables $T_m = \sqrt{P/[G(1 + \kappa a)]}\theta_m$ we shall reduce the system (2.81) to the form:

$$\Delta_2^2\psi_1 - \sqrt{R_1}\frac{\partial\theta_1}{\partial x} = 0, \quad \Delta_2\theta_1 - \sqrt{R_1}\frac{\partial\psi_1}{\partial x} = 0,$$

$$\Delta_2^2\psi_2 - \sqrt{R_1}\zeta\frac{\partial\theta_2}{\partial x} = 0, \quad \Delta_2\theta_2 - \sqrt{R_1}\chi\kappa\frac{\partial\psi_2}{\partial x} = 0, \tag{2.83}$$

$$R_1 = GP/(1 + \kappa a), \quad \zeta = \nu/\beta.$$

Boundary condititons are similar to (2.82).

The calculation range is covered by a rectangular mesh, and differential analogs of equations and boundary conditions are constructed. Denoting the vectors-columns by $\{\psi\}$ and $\{\theta\}$, the elements of which are the stream functions and temperature values at the mesh nodes, we can write the finite difference equations in a matrix form:

$$(A)\{\psi\} + \sqrt{R_1}(B)\{T\} = 0,$$
$$(C)\{T\} + \sqrt{R_1}(D)\{\psi\} = 0,$$

where (A), (B), (C) and (D) are arbitrary matrices. Eliminating $\{T\}$ we get the following problem in eigenvalues

$$[(M) - \frac{1}{R_1}(E)]\{\psi\} = 0, \tag{2.84}$$

where (E) is a unit matrix,

$$(M) = (A)^{-1}(B)(C)^{-1}(D).$$

Eigenvalues and eigenvectors of the system (2.84), starting from the largest modulus value (the smallest critical Rayleigh number), have been found on the basis of power method modification described in [41].

To apply the finite element method it is convenient to present the system (2.83) in the form[4]:

---

[4] The finite element method is also used for calculation of the thermocapillary convection threshold in enclosed vessel [188]; the problem was considered in one-layer approach.

$$\Delta_2 \psi_m - \varphi_m = 0,$$

$$\Delta_2 \varphi_m - \sqrt{R_1} e_m \frac{\partial \theta_m}{\partial x} = 0, \qquad (2.85)$$

$$\Delta_2 \theta_m - \sqrt{R_1} f_m \frac{\partial \psi_m}{\partial x} = 0 \quad (m = 1, 2),$$

where

$$e_1 = f_1 = 1, \quad e_2 = \zeta, \quad f_2 = \chi\kappa.$$

It is possible to construct a functional the extremum of which is realized at the equations solutions (2.85) with boundary conditions similar to (2.82):

$$I = I_1 + \frac{\chi}{\zeta} I_2 + \left(1 - \frac{\eta\chi}{\zeta}\right) \int_0^l \left(\varphi_1 \frac{\partial \psi_1}{\partial z}\right)_{z=0} dx,$$

$$I_m = \int\limits_{S_m} \left\{ \frac{\partial \psi_m}{\partial x} \frac{\partial \varphi_m}{\partial x} + \frac{\partial \psi_m}{\partial z} \frac{\partial \varphi_m}{\partial z} + \frac{1}{2}(\varphi_m)^2 - \right.$$

$$\frac{1}{2}\frac{e_m}{f_m}\left[\left(\frac{\partial T_m}{\partial x}\right)^2 + \left(\frac{\partial T_m}{\partial z}\right)^2\right] + \frac{1}{2}\sqrt{R_1} e_m \left(\psi_m \frac{\partial T_m}{\partial x}\right.$$

$$\left.\left. - T_m \frac{\partial \psi_m}{\partial z}\right)\right\} dx\, dz \qquad (m = 1, 2)$$

(integration is performed over the vessel part filled up with respective fluid).

In accordance with the finite element method, the section is divided into any number of triangular elements. The integrals I_1 and I_2 are presented as integral sums in squares of separate elements, and the integral over the interface is presented as the integral sum of separate sections. Functions ψ_m, φ_m and T_m inside each of such elements are approximated by polynoms of the first power in x and z, coefficients of which are expressed through the function values at the nodal points. As a result, the integral over a separate element is expressed through the variables values at the elements tops and the functional I is presented as the values function ψ_m, φ_m and T_m at the nodal points. Variation of I by these values leads to a linear algebraic equations system, where stream function, vortex and temperature values are the unknown variables in the nodal points, forming the vector-columns $\{\psi\}$, $\{\varphi\}$ and $\{T\}$. $\{\varphi\}$ and

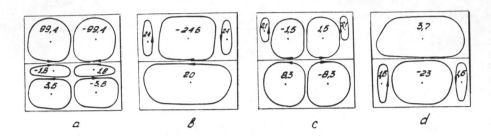

Figure 2.40. Critical motions for the system with parameters $\eta = \zeta = 0.5$; $\kappa = \chi = 1$. Critical values of parameter R_1 : 1721 (a), 1794 (b), 3227 (c), 3239 (d).

$\{T\}$ being excluded we get the problem of (2.84) form, which like in the case of the finite difference method, was solved by the power method.

As an example we shall describe the calculation results for critical motions, carried out by the finite element method. In all the calculations the ratio of layers thicknesses and vessel dimensions are assumed to be fixed: $a = 1$, $L = 2$.

Let us consider first the results corresponding to the model system with parameters: $\eta = \zeta = 0.5$, $\kappa = \chi = 1$. This case is of interest in connection with the nonlinear calculation results discussed in Chapter 4. In Fig. 2.40 the stream lines pictures for the four lower critical motions are presented. In the vortex centres the maximum-by-modulus stream function values (vortex intensities) are shown. These values can be used for the comparison of intensities within one figure limits only, presenting a certain critical motion. The intensity values in different figures are not connected with each other; the solution amplitude of uniform equation depends on the method of normalization used in calculation of the system (2.84) eigenvector. The low-intensity vortices near the interface are not shown.

"The local" Rayleigh numbers for both the media are linked by the relation $R_1 = 2R_2$ (see (2.39)). That is why the lower critical motions are characterized by the fact, that convection arises mainly in the upper fluid, while in the lower fluid a weak motion induced by tangential stresses and temperature heterogeneity on the interface takes place. The third and the fourth critical motions which have much higher threshold

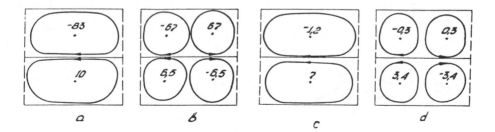

Figure 2.41. Critical motions for the system with parameters $\eta = \zeta=0.5$; $\kappa = \chi=1$ (a convective cell). Critical values of parameter R_1: 854 (a), 1030 (b), 1440 (c), 1893 (d).

Rayleigh numbers are connected with convection arising mainly in the lower fluid.

In order to find the boundary conditions effect on the lateral walls on the basis of the same method the calculations with boundary conditions

$$x = 0, L : \psi = \frac{\partial^2 \psi}{\partial x^2} = 0, \quad \frac{\partial T}{\partial x} = 0,$$

corresponding to the symmetry conditions on the convective cells boundaries in the infinite layers system were carried out. The results are presented in Fig. 2.41. Lateral boundary conditions change is seen to cause a change in both the critical Rayleigh numbers and the structures order.

In Fig. 2.42 the critical motions for a real olive oil — 85 % glycerine solution system for which the relation of local Rayleigh numbers is $R_1/R_2 \simeq 2.8$ is shown.

Let us pay attention to the real and model systems motion structure bearing some definite resemblance. To conclude this point we shall mention the work of V. V. Alekseev and A. A. Aleksandrov [3]. The paper is devoted to the case of two-dimensional convection arising in a cylindrical reservoir of rectangular section filled up with two media (air and water). Convection arises if on the interface, a constant temperature which is higher than the one on solid horizontal walls is kept. Calculations were carried out by the variational method. The dependence of the critical Rayleigh number and characteristics of critical motion on the vessel walls ratio were investigated.

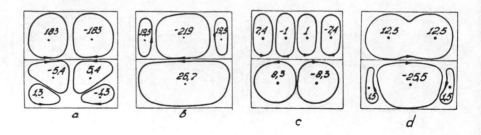

Figure 2.42. Critical motions for a real system ($\eta=0.74$, $\zeta=0.666$, $\chi=0.88$, $\kappa=0.54$). Critical values of parameter R_1 : 1777 (a), 1820 (b), 4897 (c), 4939 (d).

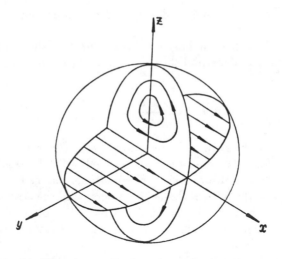

Figure 2.43. Convection in a spherical cavity; origin of coordinates is in the cavity centre.

2. Convection in a spherical vessel. Let us consider a spherical vessel of radius r_0 filled up by two immiscible fluids with horizontal interface $z = h$ (see Fig. 2.43). As earlier all the variables, concerning the upper fluid are denoted by index 1, and variables corresponding to the lower fluid by index 2. In the solid mass, surrounding the spherical vessel, the temperature distribution providing the opportunity of steady state in the system (heating from below is assumed, $\theta > 0$) is assumed:

$$z > h: \quad T_1 = \theta \frac{r_0 - z}{(r_0 - h) + (r_0 + h)(\kappa_1 \kappa_2^{-1})},$$

$$z < h: \quad T_2 = \theta \frac{(r_0 - h) - z(\kappa_1 \kappa_2^{-1})}{(r_0 - h) + (r_0 + h)(\kappa_1 \kappa_2^{-1})}.$$

Let us turn to dimensionless variables taking for units of length, time, velocity, pressure and temperature respectively r_0, r_0^2/ν_1, ν_1/r_0, $\rho_1 \nu_1^2/r_0^2$ and θ . For neutral monotonous disturbances we get

$$-e_m \nabla p_m + c_m \Delta v_m + b_m G T_m \gamma = 0,$$

$$\frac{d_m}{P} \Delta T_m = A_m(v_m \gamma) \quad (m = 1, 2),$$

where

$$A_1 = -\frac{1}{(1 - H) + \kappa(1 + H)},$$

$$A_2 = -\frac{\kappa}{(1 - H) + \kappa(1 + H)},$$

$$H = \frac{h}{l}; \quad G = \frac{g\beta_1 \theta r_0^3}{\nu_1^2};$$

the other notations are the same as in Section 2.1.

The boundary conditions on the solid suface of spherical vessel $r = 1$ have the form:

for isothermal boundaries:

$$v_m = 0; \quad T_m = 0;$$

for heat insulated boundaries:

$$v_m = 0; \quad \frac{\partial T_m}{\partial r} = 0.$$

on the interface $z = h$:

$$v_{1z} = v_{2z} = 0; \quad v_{1x} = v_{2x}; \quad v_{1y} = v_{2y};$$

$$\eta \frac{\partial v_{1x}}{\partial z} = \frac{\partial v_{2x}}{\partial z}; \quad \eta \frac{\partial v_{1y}}{\partial z} = \frac{\partial v_{2y}}{\partial z};$$

$$T_1 = T_2; \quad \kappa \frac{\partial T_1}{\partial z} = \frac{\partial T_2}{\partial z}.$$

The stated boundary value problem has been solved for a number of particular cases. In the article [47] the case $h = 0$ (the interface is on the equatorial plane) is considered; the convection threshold is determined both for isothermal and heat insulated boundaries. Let us consider the calculation method used for the vessel with heat insulated boundaries in a more detailed way. It is known [194] that for a homogeneous fluid among the critical motions, satisfying the fluid non-flow over the interface the motion of the structure shown in Fig. 2.43 has the lowest threshold Grashof number value. In [47] the critical motion of the given structure is supposed to be the most dangerous one for a two-layer system also. It follows from [194] that in the flow of the type considered, the fluid particle trajectories are close to a plane and are contained in vertical planes. This allows us to set approximately $v_y = 0$ and to introduce the stream functions ψ_m connected with velocity components by usual relations

$$v_{m,x} = \frac{\partial \psi_m}{\partial z}, \quad v_{m,z} = -\frac{\partial \psi_m}{\partial x}.$$

The problem solution was obtained by the Galerkin method with two basis functions for stream funtion and temperature:

$$\begin{pmatrix} \psi_1 \\ \psi_2 \end{pmatrix} = \alpha_1 \begin{pmatrix} \psi^{(1)} \\ \psi^{(1)} \end{pmatrix} + \alpha_2 \begin{pmatrix} \psi^{(2)} \\ \eta\psi^{(2)} \end{pmatrix},$$

$$\begin{pmatrix} T_1 \\ T_2 \end{pmatrix} = \alpha_3 \begin{pmatrix} T^{(1)} \\ \kappa T^{(1)} \end{pmatrix} + \alpha_4 \begin{pmatrix} T^{(2)} \\ T^{(2)} \end{pmatrix},$$

where

$$\psi^{(1)} = (1 - r^2)^2 z, \quad \psi^{(2)} = (1 - r^2)^2 (6z^2 - 23z^4);$$

functions $T^{(1)}$ and $T^{(2)}$ are the solutions of Poisson equation

$$\Delta T^{(1)} = -v_z^{(1)}, \quad \Delta T^{(2)} = -v_z^{(2)},$$

satisfying boundary conditions for temperature on the sphere surface.

In [158] the problem was solved at arbitrary h, but only for parameters values η, κ, χ and ζ close to unity. For that, first the stability problem of homogeneous fluid in a sphere with the condition of fluid non-flow across the plane $z = h$ has been solved; the variational method was used. The corrections stipulated by the fluids parameters difference were calculated on the basis of perturbation theory.

The conditions of convection generation in a spherical vessel filled up with two immiscible fluids with an interface in the equatorial plane were experimentally investigated in [126].

3. Thermocapillary instability of a spherical interface. Let the interface between the fluids have a spherical form $r = \mathcal{R}$ (the spherical coordinate system r, θ and φ is used). The first fluid fills the volume inside the sphere and the other — outside the sphere. Gravity is not taken in consideration; The surface tension on the interface depends on the temperature according to the law $\Sigma = \Sigma_0 - \alpha T$.

If the temperature distribution depends on r, the steady state is possible in the system. In [8] stability for the following temperature distribution

$$0 < r < R : T(r) = \theta; \quad R < r < \infty : T_2(r) = \theta\left(1 - \frac{R}{r}\right)$$

is studied. Let us note that this temperature distribution satisfies the heat conductivity equation $\Delta T_m(r) = 0$, $m = 1, 2$, but does not satisfy the heat flux balance condition on the interface: at $r = \mathcal{R}$ $\kappa dT_1/dr \neq dT_2/dr$. This circumstance may be interpreted as a consequence of heat release on the media interface.

Carring out normalization to the parameters of the first fluid and variables θ and $\mathcal{R}$, we get the following boundary value problem for small neutral disturbances ($\lambda = 0$) (the distortion of the interface is not considered):

$$c_m \Delta v_m - e_m \nabla p_m = 0; \quad \text{div} v_m = 0,$$

$$\Delta T_1 = 0; \quad \frac{1}{\chi P}\Delta T_2 - \frac{v_{2,r}}{r^2} = 0 \tag{2.86}$$

with boundary conditions

$$r \to \infty : T_1, v_1 \to 0; r = 0 : |T_2|, |v_2| < \infty; \tag{2.87}$$

$$r = 1 : \quad \eta \left(\frac{\partial v_{1,\theta}}{\partial r} - \frac{v_{1,\theta}}{r} \right) - \left(\frac{\partial v_{2,\theta}}{\partial r} - \frac{v_{2,\theta}}{r} \right) - Mr \frac{1}{r} \frac{\partial T_1}{\partial \theta} = 0,$$

$$\eta \left(\frac{\partial v_{1,\varphi}}{\partial r} - \frac{v_{1,\varphi}}{r} \right) - \left(\frac{\partial v_{2,\varphi}}{\partial r} - \frac{v_{2,\varphi}}{r} \right) - Mr \frac{1}{r \sin \theta} \frac{\partial T_1}{\partial \varphi} = 0,$$

$$v_{1,r} = v_{2,r} = 0; v_{1,\theta} = v_{2,\theta}; v_{1,\varphi} = v_{2,\varphi}; T_1 = T_2; \quad \kappa \frac{\partial T_1}{\partial r} = \frac{\partial T_2}{\partial r}. \tag{2.87}$$

Here $Mr = \alpha \theta R / \eta_2 \nu_1$; in the rest the notation is the same as in Section 2.1.

The boundary value problem (2.86), (2.87) has spherical symmetry and permits separation of variables if the solution is obtained in a series in the generalized spherical function [8]. The problem has the account number of eigenvalues

$$Mr_l = \frac{1}{\chi P} 4(2l+1)(l+1+\kappa l)(\eta+1), \quad l = 2, 3, \ldots$$

to each of which $(2l+1)$ eigenfunctions correspond. The smallest eigenvalue is realized at $l = 2$ and is equal to

$$Mr_* = \frac{20}{\chi P}(3+2\kappa)(\eta+1).$$

2.9.   Convective stability of flows

In previous paragraphs we considered systems at steady state (at small Grashof and Marangoni numbers) which the instability leads to the appearance of a convective flow. But a more complicated situation is possible if already in the reference state the fluids flow along the interface in the system.

1.   The motion of solid boundaries. In the given section the influence of solid boundaries motion on convective stability of a two-layer system at the temperature gradient directed across the layer is investigated [67–69].

Let us consider a system of two flat horizontal layers of immiscible fluids, described in Section 1.2. We shall assume that the upper and the lower solid layer boundaries perform the motion with constant velocity directed along the x axis and being equal with respect to V_1 and V_2. The profile of plane-parallel flow in the first and the second media is described by

$$U_1(z) = U_0 + Vz/a_1 \quad (0 \le z \le a_1), \tag{2.88}$$

$$U_2(z) = U_0 + Vz\eta_1/a_1\eta_2 \quad (-a_2 \le z \le 0), \tag{2.89}$$

where

$$U_0 = \left(V_2 + V_1\frac{\eta_1 a_2}{\eta_2 a_1}\right)\left(1 + \frac{\eta_1 a_2}{\eta_2 a_1}\right)^{-1},$$

$$V = (V_1 - V_2)\left(1 + \frac{\eta_1 a_2}{\eta_2 a_1}\right)^{-1}$$

Later on we shall consider the flow in the reference system moving with velocity U_0 in the direction of x axis; in this case the term U_0 in formulas (2.88), (2.89) is absent.

Let us consider a normal disturbance, characterized by any wave vector k. It is not difficult to prove that the equations describing the evolution of this disturbance include the velocity of plane-parallel flow only in projection on to the vector k, i.e. combination Vk_x/k. Velocity component perpendicular to the vector k does not influence stability. That is why it is possible to restrict ourselves with consideration of two-dimensional disturbances with $k_x = k$; the results for arbitrary orientation of the wave number are obtained by simply changing V for Vk_x/k.

Introducing the Reynolds number

$$\mathrm{Re} = \frac{Va_1}{\nu_1},$$

let us write the boundary value problem for two-dimensional normal disturbances (compare with (2.14)–(2.23)) :

$$\lambda D\psi_m - ik\,\mathrm{Re}f_m z D\psi_m = -c_m D^2\psi_m + ikGb_m T_m,$$
$$-\lambda T_m + ik\,\mathrm{Re}f_m z T_m - ik A_m \psi_m = \frac{d_m}{P}DT_m, \tag{2.90}$$

$$z = 1: \ \psi_1 = \psi_1' = T_1 = 0,$$
$$z = -a: \ \psi_2 = \psi_2' = T_2 = 0, \tag{2.91}$$

$$z = 0: \quad \psi_1''' - \frac{1}{\eta}\psi_2''' + (\lambda - 3k^2)\psi_1' + \left(-\frac{\lambda}{\rho} + \frac{3k^2}{\eta}\right)\psi_2' +$$

$$ikRe(1-v)\psi_1 + ik\left[Ga\left(\rho^{-1} - 1 + \right.\right. \tag{2.92}$$

$$\left.\left. \delta_\beta \frac{s(1 - \rho^{-1}\beta^{-1})}{1 + \kappa a}\right) + Wk^2\right]\eta = 0,$$

$$\eta(\psi_1'' + k^2\psi_1) - (\psi_2'' + k^2\psi_2) - ikMr\left(T_1 - \frac{s}{1 + \kappa a}h\right) = 0,$$

$$\psi_1' - \psi_2' = (\eta - 1)\text{Re}h,$$

$$\psi_1 = \psi_2 = -i\frac{\lambda}{k}h,$$

$$T_1 - T_2 = \frac{s(1 - \kappa)}{1 + \kappa a}h \quad \kappa T_1' - T_2' = 0.$$

Here the same notation as in Section 2.1 is used; $f_1 = 1$, $f_2 = \eta$.

The full investigation of the system (2.90) − (2.92) has not yet been carried out. Let us restrict ourselves with the consideration of longwave disturbances limit, the analysis of which can be realized analytically [67]. Using expansions in the wave number k similar to that used in Section 2.7, we get the following expression for the disturbances decrement in the longwave region:

$$\lambda = \lambda^{(1)}k + \lambda^{(2)}k^2 + \cdots,$$
$$\lambda^{(1)} = -iRe2\eta(\eta - 1)a^2(1 + a)\mathcal{Z}, \tag{2.93}$$
$$\lambda^{(2)} = C_1Ga + C_2sG + C_3sMr + C_4(\eta - 1)Re^2,$$

where

$$\mathcal{Z} = (1 + 4\eta a + 6\eta a^2 + 4\eta a^3 + \eta^2 a^4)^{-1},$$

$$C_1 = \frac{1}{3}\eta a^3 (1 + \eta a)(\rho^{-1} - 1)\mathcal{Z},$$

$$C_2 = \left\{ \frac{1}{3}\eta a^3 \frac{(1 - \rho^{-1}\beta^{-1})(1 + \eta a)}{1 + \kappa a} + \right.$$

$$\frac{(\kappa - 1)a^2}{120(1 + \kappa a)^2}[\eta(11\eta a^2 + 14a + 3) -$$

$$\left. \kappa \nu a^3 \beta^{-1}(3\eta a^2 + 14\eta a + 11)] \right\}\mathcal{Z},$$

$$C_3 = \frac{\kappa a^2 (1 + a)(1 - \eta a^2)}{2(1 + \kappa a)^2}\mathcal{Z}.$$

The expression for C_4 is extremely complicated (see [67]), that is why we shall present it for some limiting cases only:

1. At $\eta \ll 1$ $C_4 = \frac{\eta a^2}{60}$
2. At $|\eta - 1| \ll 1$

$$C_4 = -\frac{a}{60(1 + a)^6}[32(1 - \nu)a^4 + (\nu a^4 - 1)a(1 + a)^2].$$

3. At $a \ll 1$ $C_4 = \frac{\eta a^2}{60}$.

Let us remind of the fact that the formula (2.93) concerns two-dimensional disturbances with the wave vector k parallel to the x axis. For the disturbances with the arbitrary wave vector orientation in expression (2.93) the change of Re for $\mathrm{Re}k_x/k$ must be carried out.

In the cases when the influence of boundary motions is destabilizing, i.e. $C_4(\eta - 1) < 0$, the most dangerous are two-dimensional disturbancs $k_x = k$, $k_y = 0$, the decrement for which is described by the formula (2.93). If the boundary motion has a stabilizing effect (as $C_4(\eta - 1) > 0$) instability is caused by the disturbances, the wave number of which is perpendicular to the velocity of the main flow ($k_x = 0$, $k_y = 1$). For such disturbances the decrement does not depend on the parameter Re and is defined by formula (2.93) without the last term.

Let us note, that as the ratio $G/Ga = \delta_\beta$ is assumed to be small, the second term in (2.93) has to be taken into account only for fluids with close densities ($|\rho^{-1} - 1| \ll 1$).

2. Velocity jump on the interface. In the previous point the stability of independent of longitudinal coordinate plane-parallel flow (2.88), (2.89) was considered. Such flow develops at rather large distances from the entrance hole through which the fluids come into the channel. In the entrance part, however, the velocity profiles of both the fluids can be close to constant excluding the boundary layer region, the width of which is small in comparison with the fluids layers thicknesses. In this case the following flow approximation becomes adequate:

$$U_1(z) = U_1^0, 0 \le z \le \infty; U_2(z) = U_2^0, -\infty \le z \le 0.$$

The stability analysis of a two-layer flow with the velocity on the interface for the layers of infinite thickness is fulfilled in [186]. Similarly to [174], the convective case with mass transfer through the interface in the absence of the interface deformation has been considered. As mentioned in Section 2.5 this case corresponds to the change $\kappa = \chi = D$, where D is diffusion coefficients ratio in the first and the second media. The dispersion correlation which let the stability criteria be defined analytically is obtained.

CHAPTER 3

Subthreshold Convection in a System with an Interface

Let us go on to nonlinear convective regimes. The first to be considered is the case of convection in an enclosed cavity Section 3.1, for, linear problem solutions being nondegenerate, the bifurcation picture in this situation is the most simple. In the following two sections the convective regimes in a horizontal layer without an interface deformation Section 3.2 and with a deformation Section 3.3 will be considered.

3.1. Convection in an enclosed volume

1. A flat interface. We shall start with nonlinear convective regimes in an enclosed cavity D, which is filled with two immiscible fluids [108]. The fluids are supposed to be separated by a flat interface $z = 0$. The applicability of this approach have been analysed in Section 1.2. Let θ be a characteristic temperature difference in the system and a_1 a characteristic cavity dimension. Then, choosing the units as in Section 1.2, we get the system of equations (1.11)–(1.12). The temperature distribution is supposed to be kept on the solid cavity boundary Γ providing mechanical steady state in both fluids:

$$z > 0 : \quad v_1 = 0, \quad T_1 = A_1 z,$$
$$z < 0 : \quad v_2 = 0, \quad T_2 = A_2 z \quad (A_2 = \kappa A_1).$$

On the flat interface $z = 0$ we have

$$\eta \frac{\partial v_{1,x}}{\partial z} - \frac{\partial v_{2,x}}{\partial z} - Mr \frac{\partial T_1}{\partial x} = 0, \tag{3.1}$$

$$\eta \frac{\partial v_{1,y}}{\partial z} - \frac{\partial v_{2,y}}{\partial z} - Mr \frac{\partial T_1}{\partial y} = 0,$$

$$v_{1,x} = v_{2,x}, \quad v_{1,y} = v_{2,y},$$

$$v_{1,z} = v_{2,z} = 0,$$

$$T_1 = T_2,$$

$$\kappa \frac{\partial T_1}{\partial z} = \frac{\partial T_2}{\partial z}.$$

We shall count off the temperature and pressure from the values corresponding to the steady state. Then nonlinear convective equations with notation of Section 2.1 can be written in the form

$$\frac{\partial v_m}{\partial t} + (v_m \nabla) v_m = -e_m \nabla p_m + c_m \Delta v_m + b_m G T_m \gamma, \qquad (3.2)$$

$$\frac{\partial T_m}{\partial t} + (v_m \nabla) T_m + A_m (v_m \gamma) = \frac{d_m}{P} \Delta T_m, \quad \mathrm{div} v_m = 0;$$

the boundary conditions on the solid boundary Γ are

$$z > 0: \quad v_1 = 0, \quad T_1 = 0; \quad z < 0: \quad v_2 = 0, \quad T_2 = 0. \qquad (3.3)$$

The interface conditions keep the form (3.1). Problem (3.1)–(3.3) corresponds to equilibrium zero solution at any parameters. The reference stability is defined by the spectrum of the linearized problem (λ is a disturbance decrement):

$$-e_m \nabla p_m + c_m \Delta v_m + b_m G T_m \gamma = -\lambda v_m \qquad (3.4)$$

$$\frac{d_m}{P} \Delta T_m - A_m (v_m \gamma) = -\lambda T_m, \quad \mathrm{div} v_m = 0$$

with boundary conditions (3.1) and (3.3).

As shown in 2 Section 2.1, the stability problem of a two-layer system is non self-adjoint if the correlations between parameters (2.33) are not fulfilled. The adjoint problem has the form:

$$-e_m \nabla p_m^c + c_m \Delta v_m^c + b_m' G T_m^c \gamma = -\lambda v_m^c, \qquad (3.5)$$

$$\frac{d_m}{P} \Delta T_m^c - f_m' A_m (v_m^c \gamma) = -\lambda T_m^c, \qquad (3.6)$$

$$\mathrm{div} v_m^c = 0; \qquad (3.7)$$

on the solid boundary Γ:

$$z > 0: \quad v_1^c = 0, \quad T_1^c = 0;$$
$$z < 0: \quad v_2^c = 0, \quad T_2^c = 0; \tag{3.8}$$

on the interface $z = 0$:

$$\frac{\partial v_{1,x}^c}{\partial z} = \frac{\partial v_{2,x}^c}{\partial z}, \quad \eta \frac{\partial v_{1,y}^c}{\partial z} = \frac{\partial v_{2,y}^c}{\partial z}, \tag{3.9}$$

$$v_{1,x}^c = v_{2,x}^c \quad v_{1,y}^c = v_{2,y}^c, \quad v_{1,z}^c = v_{2,z}^c = 0, \tag{3.10}$$

$$\kappa \frac{\partial T_1^c}{\partial z} + \frac{M r A_1 P}{\eta G} \frac{\partial v_{2,z}^c}{\partial z} = \frac{\partial T_2}{\partial z}. \tag{3.11}$$

Here $b_1' = f_1' = 1, b_2' = \rho \chi, f_2' = (\beta \rho \chi)^{-1}$.
The condition of the solvability of a heterogeneous linear problem

$$-e_m \nabla p_m + c_m \Delta v_m + b_m G T_m \gamma = f_m,$$
$$\frac{d_m}{P} \Delta T_m - A_m(v_m \gamma) = g_m, \quad \text{div} v_m = 0 \tag{3.12}$$

with boundary conditions (3.1), (3.3) can be written in the form:

$$\int_{z>0} dV \left(v_1^c f_1 - T_1^c g_1 \frac{G}{A_1} \right) + \int_{z<0} dV \left(\frac{1}{\rho} v_2^c f_2 - T_2^c g_2 \frac{G \chi}{A_2} \right) = 0. \tag{3.13}$$

At $G \to 0$ after rescaling $T_m^{c'} = G T_m^c$ equations (3.5)–(3.6) and (3.11) get the form (the prime is omitted):

$$-e_m \nabla p_m^c + c_m \Delta v_m^c + b_m' T_m^c \gamma = -\lambda v_m^c, \tag{3.14}$$

$$\frac{d_m}{P} \Delta T_m^c = -\lambda T_m^c, \tag{3.15}$$

$$z = 0: \quad \kappa \frac{\partial T_1^c}{\partial z} + \frac{M r A_1 P}{\eta} \frac{\partial v_{2,z}^c}{\partial z} = \frac{\partial T_2^c}{\partial z}. \tag{3.16}$$

In this case boundary conditions (3.8) – (3.10) are not changed.

As shown in Chapter 2, the instability of a two-layer system may be of a monotonous or oscillatory character. First we shall consider the case of a monotonous instability. Let at $G = G^{(0)}$ and $Mr = Mr^{(0)}$ the linear problem (3.4), (3.1) and (3.3) has nondegenerate eigenvalue $\lambda = 0$ which corresponds to the real eigenfunction $U^{(1)} = (v^{(1)}, T_1^{(1)}, p^{(1)})$. We search for solution $U(v, T, p)$ of problem (3.1)–(3.3) at a small deflection of G from $G^{(0)}$ as a series in a small parameter:

$$U = \varepsilon U^{(1)} + \varepsilon^2 U^{(2)} + ..., \quad G = G^{(0)} + \varepsilon G^{(1)} + \varepsilon^2 G^{(2)} + ..., \quad (3.17)$$

(the case of deviation of Mr from $Mr^{(0)}$ is considered separately).

Substituting expansions (3.17) in (3.1)–(3.3) in the first order in ε we get a homogeneous linear problem which determins $G^{(0)}$ and $U^{(1)}$, and in the second order in ε we get a heterogeneous problem:

$$-e_m \nabla p_m^{(2)} + c_m \Delta v_m^{(2)} + b_m G^{(0)} T_m \gamma =$$

$$- b_m G^{(1)} T_m^{(1)} \gamma + (v_m^{(1)} \nabla) v_m^{(1)}, \qquad (3.18)$$

$$\frac{d_m}{P} \Delta T_m - A_m(v_m \gamma) = v_m^{(1)} \nabla T_m^{(1)}, \quad \mathrm{div} v_m^{(2)} = 0.$$

The boundary conditions are homogeneous and are not quoted. According to (3.13), the solvability condition of this problem determines the variable $G^{(1)}$:

$$G^{(1)} = \frac{I}{\mathcal{K}}, \qquad (3.19)$$

$$I = \int\limits_{z>0} dV \left[v_1^c(v_1^{(1)} \nabla) v_1^{(1)} - \frac{G^{(0)}}{A_1} T_1^c v_1^{(1)} \nabla T_1^{(1)} \right] +$$

$$\int\limits_{z<0} dV \left[\frac{1}{\rho} v_2^c(v_2^{(1)} \nabla) v_2^{(1)} - \frac{G^{(0)} \chi}{A_2} T_2^c v_2^{(1)} \nabla T_2^{(1)} \right], \qquad (3.20)$$

$$\mathcal{K} = \int\limits_{z>0} dV (v_1 \gamma) T_1^{(1)} + \int\limits_{z<0} dV \frac{1}{\beta} v_2^c \gamma T_2^{(1)}. \qquad (3.21)$$

It is not difficult to show that variable $\mathcal{K}$ is proportional to $(d\lambda/dG)_{G=G^{(0)}}$ and , respectively, differs from zero. For a homogeneous fluid $I = 0$ owing to the adjoint problem ($v^c = v^{(1)}$, $T^c = T^{(1)}$), and identities

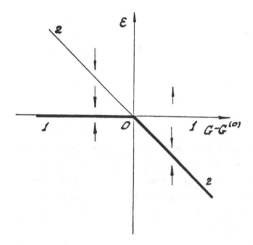

Figure 3.1. Branching of convective motion (line 2) from the steady state (line 1) in the case of bilateral bifurcation ($G^{(1)} < 0$). Thick lines relate to the stable state, thin lines — to unstable one.

$$\int dV v^{(1)}(v^{(1)}\nabla)v^{(1)} = 0, \quad \int dV T^{(1)} v^{(1)} \nabla T^{(1)} = 0,$$

which are fulfilled for any fields $v^{(1)}$ and $T^{(1)}$ satisfying the condition $\mathrm{div} v^{(1)} = 0$ in the volume, and the vanishing conditions of the normal velocity component and temperature on the integration region boundary. Generally speaking, for a two-layer system the variable $G^{(1)}$ differs from zero, this corresponds to the so-called two-hand side bifurcation [36] (see Fig. 3.1). Solution corresponding to the convective motion (line 2) and solution corresponding to the reference state (line 1) exist both in supercritical region ($G > G^{(0)}$) and in subcritical region $G < G^{(0)}$. Passing through value $G = G^{(0)}$ only the stability character of both solutions in relation to infinitely small disturbances changes: at $G < G^{(0)}$ the reference state is stable, and at $G > G^{(0)}$ the convective motion is stable.

In region $G < G^{(0)}$ the steady state is unstable with respect to disturbances whose amplitude ε is larger than the amplitude of an unstable stationary motion: $\varepsilon > (G - G^{(0)})/G^{(1)}$. Nonlinear disturbance increase finally leads to finite-amplitude convection, the description of which lies, however, outside the small parameter applicability.

There are, however, situations for which variable $G^{(1)}$ vanishes. For example, this happens if the fluid parameters satisfy the adjoint conditions of problem (2.33). Besides, $G^{(1)}$ can vanish owing to the specific symmetry properties of the main and adjoint problem solutions. For example, for two-dimensional convection in a horizontal cylinder of a rectangular cross section D $(0 \leq x \leq L, -\infty < y < \infty, -a \leq z \leq 1)$, the solutions of the linear problem for monotonous disturbances, studied in Section 2.8, disintegrate into two classes for which:

1. v_{zm}, T_m are non-even functions, v_{xm} is an even function on $x - L/2$;
2. v_{zm}, T_m are even functions, v_{xm} is an odd function on $x - L/2$.

The solutions of an adjoint problem possess the same symmetry properties. It is not difficult to prove that for the solutions of the first class we have $G^{(1)} = 0$, and for the solutions of the second class in general case we have $G^{(1)} \neq 0^1$.

If $G^{(1)} = 0$, then from the solvability conditions of the problem in the second order in ε it is necessary to find the value of $G^{(2)}$. The amplitude of a convective motion close to the threshold $G = G^{(0)}$ is described by

$$\varepsilon = \pm \left(\frac{G - G^{(0)}}{G^{(2)}} \right)^{1/2} \tag{3.22}$$

If $G^{(2)} > 0$, the solutions bifurcate to a supercritical region, as $G^{(2)} < 0$, the solutions bifurcate to a subcritical region (see Fig. 3.2). In the first case the "soft" bifurcation of steady convective motions takes place; in the second case we have a "hard" finite-amplitude instability of the reference state.

Let now convection arise at the deflection of Mr on $Mr^{(0)}$. Introducing expansion

$$Mr = Mr^{(0)} + \varepsilon Mr^{(1)} + \varepsilon^2 Mr^{(2)} + ...,$$

from the solvability condition in the second order in ε we get

$$Mr^{(1)} = \frac{I}{\mathcal{L}}, \tag{3.23}$$

where I is defined by (3.20),

---

[1] Similar results are obtained for thermocapillary convection in a one-layer system [188].

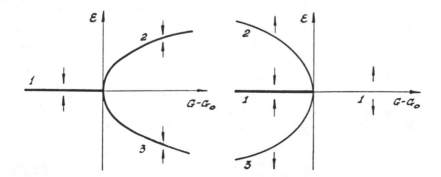

Figure 3.2. Branching of solutions in the case of supercritical (a) and sub-critical (b) unilateral bifurcation. Line 1 — the steady state, lines 2,3 — convective motions.

$$\mathcal{L} = \eta \int_{z=0} dx dy T_1 \frac{\partial v_{z,1}^c}{\partial z}. \tag{3.24}$$

As $G^{(0)} = 0$ equality (3.23) remains the same, it is necessary, however, to consider problem (3.14)–(3.16) as self-adjoint and in the expression for I we change the multiplier $G^{(0)}$ into unit.

Let us discuss the character of bifurcation for the case of oscillatory instability $(\lambda = \pm i\omega)$. We shall restrict ourselves with the investigation of bifurcation at the deflection of G on $G^{(0)}$ (the deflection of Mr on $Mr^{(0)}$ is studied similarly). We construct the solution as an expansion

$$U = \varepsilon(U^{(1)}e^{i\omega t} + U^{(1)}F^*e^{-i\omega t})+$$

$$\varepsilon^2 \sum_{l=-2}^{2} U_l^{(2)}e^{il\omega t} + .., \quad U_{-l}^{(n)} = U_l^{(n)*},$$

$$\omega = \omega^{(0)} + \varepsilon\omega^{(1)} + \varepsilon^2\omega^{(2)} + ..., \quad G = G^{(0)} + \varepsilon G^{(1)} + \varepsilon^2 G^{(2)} + ..,$$

where l and n are the integers.

In the first order in ε we obtain the linear problem, defining the critical value of parameter $G^{(0)}$, frequency $\omega^{(0)}$ and function $U^{(1)}$. In the second order in ε nonlinear terms contribute only to the right-hand side parts of the equations for $U_m^{(2)}$ with even m which always have solutions. From the solvability conditions of the boundary value problem for $U_1^{(2)}$ we find $G^{(1)} = \omega^{(1)} = 0$. It is not difficult to show by induction, that $G^{(2n+1)} = \omega^{(2n+1)} = 0$, $n = 0, 1, ...$, i.e., $G = G(\varepsilon^2)$, $\omega = \omega(\varepsilon^2)$. This is also clear from the fact that $U(t; -\varepsilon) = U(t + \pi/\omega; \varepsilon)$. Thus, the bifurcation of oscillatory solutions is always one-hand side, which agrees with general theory [36].

2. A deflected interface. In an enclosed cavity the interface between media, generally speaking, is not flat. As mentioned in Section 1.2, the form of interface $z = h(x, y)$ is defined by the normal stress balance of hydrostatic and capillary nature and can be determined from equation

$$(\rho^{-1} - 1)h + \frac{r_c^2}{\mathcal{R}} = C, \qquad (3.25)$$

where $\mathcal{R}$ is a radius of curvature, $r_c = (\sigma_0/\rho_1 g a_1^2)^{1/2}$ is a dimensionless capillary radius, C is a pressure difference on both sides of the interface. Equation (3.25) is solved with the boundary conditions on the solid surface, corresponding to the given contact angle. If the dimensionless capillary radius is small, then the interface deflection is not large and occurs near solid walls.

In the present part the influence of a small stationary interface deformation on the solution bifurcation at $G \simeq G^{(0)}$ is investigated.

Let the interface be defined by $z = h(x, y)$, where δ is a small parameter. As before, we shall count off the temperature and pressure from values $T_m^{(0)}$, $p_m^{(0)}$ which would be realized in the case of flat interface $z = 0$ and a mechanical steady state. Boundary conditions (3.1) are then changed in the following manner:

$$z = \delta h(x, y) : \quad \eta \frac{\partial v_{1,n}}{\partial n} - \frac{\partial v_{2,n}}{\partial n} - Mr\nabla_\tau(T_1 + T_1^{(0)}) = 0, \qquad (3.26)$$

$$v_{1,\tau} = v_{2,\tau},$$

$$v_{1,n} = v_{2,n},$$

$$T_1 + T_1^{(0)} = T_2 + T_2^{(0)},$$

$$\kappa \frac{\partial(T_1 + T_1^{(0)})}{\partial n} = \frac{\partial(T_2 + T_2^{(0)})}{\partial n},$$

where

$$v_{m,n} = n v_m, \quad v_{m,\tau} = v_m - n(n v_m),$$

$$\frac{\partial}{\partial n} = n\nabla, \quad \nabla_\tau = \nabla - n(n\nabla),$$

n is a normal vector to the interface. If δ is small the boundary conditions can be transfered to plane $z = 0$ by the exchange in (3.26) of the function values on the boundary by the expansion in Taylor's series:

$$f(\delta h) = f(0) + \delta h \frac{\partial f(0)}{\partial z} + \dots$$

Already at any small values of G and Mr the steady state in a fluid is impossible. Actually, the interface deflection leads to the break of the temperature constancy on the interface, which stipulates thermocapillary convection at $Mr \neq 0$. If $\kappa \neq 1$, the deflection of the interface leads also to the break of the equilibrium temperature distribution and to the onset of thermogravitational convection in case $G \neq 0$.

When $\delta \ll 1$ U can be constructed as an expansion in δ. With the approach of parameters G and Mr to the values $G^{(0)}$ and $Mr^{(0)}$, corresponding to the threshold of monotonous instability for a flat interface, the terms of expansion in δ go to infinity. The divergence of the expansion in δ is stipulated by the destruction of correlation $U = O(\delta)$ in the bifurcation point vicinity, where the solutions not going to zero at $\delta \to 0$ exist.

At $\delta = 0$ the bifurcation appears to be two-hand side. For the investigating bifurcation at $\delta \neq 0$ we assume

$$G = G^{(0)} + \varepsilon G^{(1)} \tag{3.27}$$

(case $Mr \neq Mr^{(0)}$ is investigated similarly). Following [185, 29], we choose

$$U = \sum_{n=1}^{\infty} \varepsilon^n U^{(n)}, \quad \delta = \sum_{n=1}^{\infty} \varepsilon^n \delta^{(n)} \tag{3.28}$$

and substitute (3.28) into (3.2), (3.3), (3.26).

Varied parameter $G^{(1)}$ describes the deviation of G from $G^{(0)}$; the dependence of amplitude ε on G and δ is determined by the second correlation in (3.28).

In the first order in ε we get a linear problem for $U^{(1)}$. The conditions on the interface are heterogeneous and can be transformed into the form:

$$z = 0: \quad v_{1,z}^{(1)} = v_{2,z}^{(1)} = 0, \quad v_{1,x}^{(1)} = v_{2,x}^{(1)}, \quad v_{1,y}^{(1)} = v_{2,y}^{(1)},$$

$$\eta \frac{\partial v_{1,x}^{(1)}}{z} - \frac{\partial v_{2,x}^{(1)}}{\partial z} - Mr^{(0)} \frac{\partial T_1^{(1)}}{\partial x} = -\delta^{(1)} Mr^{(0)} A_1 \frac{\partial H}{\partial x},$$

$$\eta \frac{\partial v_{1,y}^{(1)}}{\partial z} - \frac{\partial v_{2,y}^{(1)}}{\partial z} - Mr^{(0)} \frac{\partial T_1^{(1)}}{\partial y} = -\delta^{(1)} Mr^{(0)} A_1 \frac{\partial H}{\partial y},$$

$$T_1^{(1)} - T_2^{(1)} = (1 - \kappa) A_1 \delta^{(1)} H, \quad \kappa \frac{\partial T_1^{(1)}}{\partial z} = \frac{\partial T_2^{(1)}}{\partial z}.$$

The solvability condition of this problem has the form

$$\delta^{(1)} N = 0, \tag{3.29}$$

where

$$N = \int_{z=0} dSH \left(-\frac{Mr^{(0)} A_1}{\eta} \frac{\partial v_{1,z}^c}{\partial z} - \frac{G^{(0)}}{P} \frac{(1-\kappa)}{\kappa} \frac{\partial T_1^c}{\partial z} \right) \tag{3.30}$$

if $N = 0$ (for example, owing to the function symmetry properties), then $\delta^{(1)} \neq 0$, i.e. $\varepsilon \sim \delta$. In this case expansions in δ have no singularities in the first order near $R = R^{(0)}$, and the bifurcation has qualitatively the same character, as $\delta = 0$ (coefficients $G^{(n)}$ in the expansion (3.17) depend on δ). If $N \neq 0$, then (3.29) leads to $\delta^{(1)} = 0$. Then the solvability condition of the problem in the second order in ε leads to

$$-\mathcal{K}G^{(1)} + I + N\delta^{(2)} = 0, \tag{3.31}$$

where I, $\mathcal{K}$ and N are defined by (3.20), (3.21) and (3.30). Multiplying (3.30) by ε^2, we get the equation of bifurcation

$$-\mathcal{K}(G - G^{(0)})\varepsilon + I\varepsilon^2 + N\delta = 0.$$

Fig. 3.3 shows the character of solution bifurcation as function of the sign of the value $Q = N\delta/I$ for case $I/\mathcal{K} < 0$ (notation is the same as in Fig. 3.1). In case $Q > 0$ with increasing G at first disappearance and then appearance of stationary solutions take place; if $Q < 0$, in the vicinity of the point $G = G^{(0)}$ no bifurcation occurs.

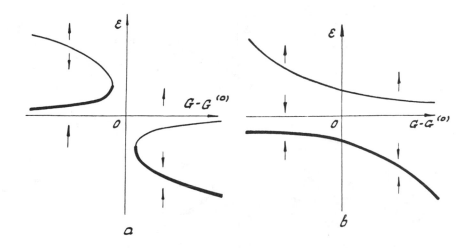

Figure 3.3. Branching of solutions in the case of bilateral bifurcation as $N\delta/I > 0$ (a) and $N\delta/I < 0$ (b).

We shall consider now the case when for a flat interface the bifurcation is two-hand side, i.e. $I = 0$. Then, choosing

$$G = G^{(0)} + G^{(2)}\varepsilon^2,$$

we can come to the equation of bifurcation

$$(G - G^{(0)} - G^{(2)}\varepsilon^2)\varepsilon - \frac{N}{K}\delta = 0,$$

where $G^{(2)}$ is the same coefficient as in (3.22). The bifurcation picture which is completely similar for the case of homogeneous fluid, obtained in [29], is shown in Fig. 3.4 ($G^{(2)} > 0$, $N\delta/K < 0$).

For an oscillatory instability the stationary interface deflection does not qualitatively change the bifurcation picture and brings to the formation of dependence of coefficients $G^{(n)}$ and $w^{(n)}$ on δ.

3.2. Convection in a system of horizontal layers

Let us consider now a nonlinear convection in a system of infinite horizontal layers. Neutral curves of mechanical equilibrium stability calculated in Chapter 2 are at the same time bifurcation boundaries solutions, corresponding to nonlinear convective regimes. The specific feature of

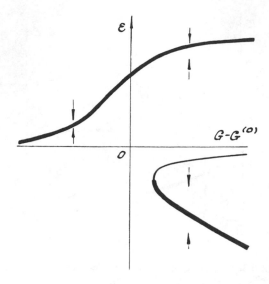

Figure 3.4. Branching of solutions as $\delta \neq 0$ in the case of supercritical unilateral bifurcation.

the problem on the generation of convection in horizontal layers is its isotropy leading to the infinite degeneration of the linearized problem spectrum: decrement λ depends only on the modulus, but not on the direction of vector k. This leads to the possibility of the flow regimes appearance, possessing different spatial structure: flows in the form of rolls, square, rectangular, hexagonal cells and so on. The selection of experimentally realized flow structure can be done on the basis of solution stability analysis of a nonlinear problem.

To be definite, later on we shall speak about the thermogravitational convection described by the nonlinear boundary problem (3.1)–(3.3). The same arguments are also true in a general case to which system (1.11)–(1.20) corresponds. Let us first consider the case of monotonous instability. We shall construct solutions of nonlinear problem (3.1)–(3.3) as an expansion in small parameter ε (the amplitude of a convective flow):

$$U = \varepsilon U^{(1)} + \varepsilon^2 U^{(2)} + \cdots, \quad G = G^{(0)} + \varepsilon G^{(1)} + \cdots \quad (3.32)$$

here $U = (v_m, p_m, T_m)$ is a vector, characterizing the motion;

$$U^{(1)}(r, z) = \sum_i \left(u_i(z)e^{ik_i r} + u_i^*(z)e^{-ik_i r} \right) \tag{3.33}$$

$r = (x, y)$; $u_i(z)$ are eigenfunctions of the linearized problem, corresponding to the disturbances with wave vector k_i; $G^{(0)}$ is the boundary of a linear stability. Substituting expansions (3.33) into equations and boundary conditions (3.1)–(3.3), one can obtain from the problem solvability conditions the expansion coefficients $G^{(n)}$ (for a more detailed variant see [152]). For a one-layer problem owing to certain symmetry properties all coefficients $G^{(n)}$ with odd n appear to be equal to zero. In the convection problem for a two-layer system coefficient $G^{(1)}$ appears to be equal to zero only for the structures, for which Fourier-expansion does not have three vectors $\{k_i\}$, having the identical modulus and forming with each other angle $120°$ (i.e. for the structures of the round, square and rectangular type). For a hexagonal structure coefficient $G^{(1)}$ in the convection problem for a two-layer system, as a rule, differs from zero, which corresponds to two-hand side bifurcation (Fig. 3.1.). The solution corresponding to the flow of hexagonal structure (line 2) and the solution corresponding to the reference state (line 1), exist both in a supercritical region ($G > G^{(0)}$), and in a subcritical region ($G < G^{(0)}$). The analysis of stability shows that unlike the enclosed cavity case, the small steady solutions of a hexagonal structure are always unstable.

At $G > G^{(0)}$ the steady state is unstable with respect to infinitesimal disturbances, and at $G < G^{(0)}$ is unstable with respect to disturbances, the amplitude of which ε is larger than the amplitude of an unstable steady motion: $\varepsilon(G - G^{(0)})/G^{(1)}$. A nonlinear increase of disturbance must finally lead to establishment of a finite-amplitude hexagonal convection, the asymptotic methods for its analysis being inapplicable. In some cases, the value of coefficient $G^{(1)}$ for a hexagonal structure may turn out to be small: $|G^{(1)}| \ll G^{(2)}$. Then the amplitude of convective flows in near-threshold region also turns out to be small, so the analysis can be continued. So far as a similar investigation for two-layer flows has not been in fact done, we shall restrict ourselves with some observations. For the first time the motion stability of a different space structure in a horizontal layer as $|G^{(1)}| \ll G^{(2)}$ on the example of convection in a fluid, the parameters of which depend on temperature has been investigated in [24]. For the convection in a layer with a free surface similar investigations (in one-layer statement) have been started in [91, 28, 31,142]. The hexagonal structure of near-threshold motions in the system with interface is known from experiment [87, 88] and nonlinear calculations [151].

Let us discuss now the case of an oscillatory instability, which leads to the development of nonsteady convective regimes. Near the threshold of stability loss with respect to the disturbances with wave number k and frequency ω_0 the establishment of convective motion may be described with the help of expansion

$$
\begin{aligned}
U = \varepsilon \sum_{i=1}^{N} \Big[& A_{i,+}(\tau) u_{i,+}(z) e^{i(\omega t + k_i r)} \\
& + A_{i,+}^* \tau u_{i,+}^*(z) e^{-i(\omega t + k_i r)} \\
& + A_{i,-}(\tau) u_{i,-}(z) e^{i(\omega t - k_i r)} \\
& + A_{i,-}^*(\tau) u_{i,-}^*(z) e^{-i(\omega t - k_i r)} \Big] \\
& + \varepsilon^2 U^{(2)} + \varepsilon^3 U(3) + \cdots,
\end{aligned}
$$

$$
\omega = \omega^{(0)} + \varepsilon^2 \omega^{(2)} + \cdots; \quad G = G^{(0)} + \varepsilon^2 G^{(2)} + \cdots, \quad \tau = \varepsilon^2 t.
$$

(3.34)

Here $\{k_i\}$, $i = 1, ..., N$ is a set of wave vectors determining the space flow structure, $A_{i,\pm}$ are complex amplitudes. Later on for simplicity we assume all $|k_i|$ to be equal. Substituting (3.34) into equations (3.1)–(3.3) from the problem solvability condition in the third order in ε it is possible to get the amplitude equation system [147]:

$$
\frac{dA_{i,+}}{d\tau} = \left(G^{(2)} + i\omega^{(2)} \right) A_{i,+} - A_{i,+} \Big(\sum_{j=1}^{N} T_{i,j} |A_{j,+}|^2 +
$$

$$
+ \sum_{j=1}^{N} S_{i,j} |A_{j,-}|^2 \Big) - A_{i,-}^* \sum_{\substack{j=1 \\ j \neq i}}^{N} U_{i,j} A_{j,+} A_{j,-};
$$

$$
\frac{dA_{i,-}}{d\tau} = \left(G^{(2)} + i\omega^{(2)} \right) A_{i,-} - A_{i,-} \Big(\sum_{j=1}^{N} T_{i,j} |A_{j,-}|^2 +
$$

$$
+ \sum_{j=1}^{N} S_{i,j} |A_{j,+}|^2 \Big) - A_{i,+}^* \sum_{\substack{j=1 \\ j \neq i}}^{N} U_{i,j} A_{j,-}^* A_{j,+}, \quad i = 1, 2, ..., N.
$$

(3.35)

Here coefficients $T_{i,j}$, $S_{i,j}$ and $U_{i,j}$ depend on the angle between wave vectors k_i, k_j only.

As $N = 1$ (roll type structures) the analysis of steady system solutions (3.35) and their stability is relatively simple. It is easy to make sure that there can be only two essentially different flow regimes: travelling waves $(A_{i,+} \neq 0, A_{i,-} = 0$ or $A_{i,-} \neq 0, A_{i,+} = 0)$ and stationary oscillations $(|A_{i,+}| = |A_{i,-}| \neq 0)$. The solutions bifurcating into a supercritical region as $T_{11} > 0$ and into a subcritical region as $T_{11} < 0$ correspond to the first type motions; the condition of their stability is $O < \mathrm{Re}T_{11} < \mathrm{Re}S_{11}$. The second type motion takes place in a supercritical region as $\mathrm{Re}(T_{11} + S_{11}) > 0$ and in a subcritical region as $\mathrm{Re}(T_{11} + S_{11}) < 0$; this type is stable if $\mathrm{Re}(T_{11} + S_{11}) > 0$, $\mathrm{Re}T_{11} > \mathrm{Re}S_{11}$.

The number of different solution types of system (3.35) grows with increasing N. For the case when $N = 3$ and vectors $\{k_i\}$ form the angle of $120°$ it is stated in [147] that there are eleven qualitatively different motion types:

1. Stationary rolls:

$$A_{1,+} = A_{1,-}; \quad A_{2,+} = A_{2,-} = A_{3,+} = A_{3,-} = 0.$$

2. Stationary rectangles:

$$A_{1,+} = A_{2,+} = A_{1,-} = A_{2,-}; \quad A_{3,+} = A_{3,-} = 0.$$

3. Stationary hexagons:

$$A_{1,+} = A_{1,-} = A_{2,+} = A_{2,-} = A_{3,+} = A_{3,-}.$$

4. Stationary regular triangles

$$A_{1,+} = A_{2,+} = A_{3,+} = -A_{1,-} = -A_{2,-} = -A_{3,-}.$$

5. Travelling rolls:

$$A_{1,-} \neq 0; \quad A_{1,+} = A_{2,+} = A_{2,-} = A_{3,+} = A_{3,-} = 0.$$

6. Rectangles travelling along a long side:

$$A_{1,-} = A_{3,-}; \quad A_{2,-} = A_{1,+} = A_{2,+} = A_{3,+} = 0.$$

7. Rectangles travelling along a short side:

$$A_{1,-} = A_{3,+}; \quad A_{2,-} = A_{3,-} = A_{1,+} = A_{2,+} = 0.$$

8. Oscillating triangles:

$$A_{1,-} = A_{2,-} = A_{3,-}; \quad A_{1,+} = A_{2,+} = A_{3,+} = 0.$$

9. Inter-roll and rectangle reconstructions:

$$A_{1,-} = A_{3,-} = A_{1,+} = -A_{3,+}; \quad A_{2,+} = A_{2,-} = 0.$$

10. Inter-rectangle reconstructions:

$$A_{2,-} = e^{i\frac{2\pi}{3}} A_{1,-}; \quad A_{3,-} = e^{i\frac{4\pi}{3}} A_{1,-}; \quad A_{1,+} = -A_{i,-} \quad (i = 1, 2, 3).$$

11. Oscillating "twisted" rectangles:

$$A_{2,-} = e^{i\frac{2\pi}{3}} A_{1,-}; \quad A_{3,-} = e^{i\frac{4\pi}{3}} A_{1,-}; \quad A_{i,+} = A_{i,-} \quad (i = 1, 2, 3).$$

For each structure type in [147] the disturbance spectrum has been found. It gives an opportunity of getting stability criteria. It turned out that for the structures of types 3, 4, 6, 7, 10 the application of equations (3.35) results in some neutral disturbance form which is not connected with original problem symmerty. The calculation of this disturbance increment required the record of amplitude equations up to the fifth order of smallness. In [138] an attempt of applying the theory developed in [147] for thermogravitational convective motions in a two-layer system caused by oscillatory instability has been made. The amplitude equations coefficients have been caculated for some concrete correlations between liquid parameters. But all bifurcating stuctures turned out to be unstable in all investigated cases.

3.3.   Thermocapillary convection with a deformable interface

As shown in Sections 1.2 and 2.7, for a longwave thermocapillary convection the deformation of the interface is an important factor which can not be ignored. From expression (2.77) it follows that at certain correlations between the parameters of the system, the longwave instability may be found most dangerous, i.e. the minimum of the neutral curve may be reached at $k = 0$. Nonlinear development of longwave disturbances possesses some peculiarities which are discussed in this section.

Nonlinear equations, determining longwave thermocapillary motions in a two-layer system, were discussed for the first time in [10]. When driving the equations the typical horizontal motion scale k^{-1} was supposed to be larger then the layer thickness. Moreover, the following correlations between the system parameters and the motion scale were supposed to be fulfilled:

$$W \sim k^{-3}, \quad Mr \sim k^{-1}, \quad Ga\delta \sim k^{-1}.$$

The amount of the interface deflection in this case turned out to be of a layer thickness order. This circumstance leads to rather complicated form modifying equations. The motion is described by two independent functions $h(x, y, t)$ and $P_1(x, y, t)$, corresponding to the interface deflection from the middle position and the pressure distribution in one of the fluids. These functions submit to the system of two essentially nonlinear equations, possessing the time derivatives only from function h. In the article [10] it is stated that the critical values of Marangoni number $Mr(k_x, k_y)$ are bifurcating points of the solutions, corresponding to the stationary space-periodic secondary motions. The complexity of the equations system being used, however, does not permit the construction of analytical solutions and also does not give an opportunity to find out, whether the bifurcation of these solutions is "soft" or "hard".

In this paragraph we shall restrict ourselves with the investigation of a longwave thermocapillary convection in a direct vicinity of the threshold Marangoni number.

Let us first describe qualitatively the nonlinear evolution of a long-wave interface deflection. It is evident, that this deflection of the interface $h(x)$ (in dimensionless variables) is equivalent to the local change of the upper and lower layers thicknesses:

$$a_1' = a_1(1 - h), \quad a_2' = a_2 + ha_1.$$

The development of convective instability in the presence of an interface deflection is defined by parameters:

$$Mr' = \frac{\alpha \theta a_1'}{\eta_2 \nu_1}, \quad Ga' = \frac{ga_1'^3}{\nu_1^2}, \quad a' = \frac{a_2'}{a_1'},$$

moreover, according to (2.75), the threshold of convection is

$$Mr_c' = \frac{2s\eta Ga'\delta}{3\kappa} \frac{(1 + \eta a')(1 + \kappa a')^2 a'}{(1 + a')(1 - \eta a'^2)}. \tag{3.36}$$

Returning to variables Mr, Ga and δ, we shall write (3.36) in the form

$$Mr_c = \frac{2s\eta Ga\delta}{3\kappa} \frac{(1 - h)[(1 - h) + \eta(a + h)][(1 - h) + \kappa(a + h)]^2(a + h)}{(1 + a)[(1 - h)^2 - \eta(a + h)^2]}. \tag{3.37}$$

Parameter h can be varied within $-a < h < 1$.
We shall give the expressions for variables

$$K_1 = \left(\frac{dMr_c}{dh}\right)_{h=0}$$

and

$$K_2 = \left(\frac{d^2Mr_c}{dh^2}\right)_{h=0},$$

defining the character of changes in the convection threshold at the interface deflection:

$$K_1 = \left[\frac{\eta(a+1)}{1+\eta a} + \frac{2\kappa(a+1)}{1+\kappa a} + \frac{1-2a}{a} + \frac{2\eta a(a+1)}{1-\eta a^2}\right] Mr_c, \quad (3.38)$$

$$
\begin{aligned}
K_2 = \Bigg[&-\frac{\eta-1}{1+\eta a} - \frac{\kappa-1}{1+\kappa a} - \frac{1}{a} - \frac{1+2\eta a}{1-\eta a^2} \\
&+ \frac{(\eta-1)(\kappa-1)}{(1+\eta a)(1+\kappa a)} + \frac{\eta-1}{a(1+\eta a)} + \frac{(\eta-1)(1+2\eta a)}{(1-\eta a^2)(1+\eta a)} \\
&+ \frac{\kappa-1}{a(1+\kappa a)} \frac{(\kappa-1)(1+2\eta a)}{(1+\kappa a)(1-\eta a^2)} + \frac{1+2\eta a}{a(1-\eta a^2)} \\
&+ \frac{\eta}{1-\eta a^2} + \left(\frac{1+2\eta a}{1-\eta a^2}\right)^2 \Bigg] Mr_c.
\end{aligned}
\quad (3.39)
$$

Depending on the values of parameters κ and η coefficient K_1 with changing a is either sign-constant or changes the sign twice. The range of parameters κ and η, in which sign $K_1(a)$ changes, consists of two separate parts. One of these numerically calculated subregions, is shadowed in Fig. 3.5 ($\kappa_m = 2/3$, $\eta_m \simeq 0.1593$); the other subregion is obtained by the transformation of $\kappa \to 1/\kappa, \eta \to 1/\eta$. If the parameters κ and η are situated in the shaded region, the dependence of sMr_c on the ratio of layers thicknesses a' is non-monotonous (see Fig. 3.6, line 1); at $a = a'_{\pm}$ coefficient K_1 vanishes. It is easy to understand, that at point a'_{-} the coefficient K_2 is negative, and at point a'_{+} it is positive. On the boundary of the shaded region the dependence sMr_c has a twist-point (line 2), in which K_1 and K_2 simultaneously vanish. Outside the shaded region sMr does not have extremums and $K_1 \neq 0$ at any a.

Let number sMr be equal to the critical value, determined by formula (2.75) for the given correlation of the layers thicknesses a. We shall make the longwave deformation of the interface. Owing to the both fluids volume conservation, the equation is fulfilled

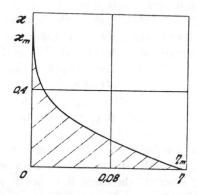

Figure 3.5. The region of parameter K_1 variety.

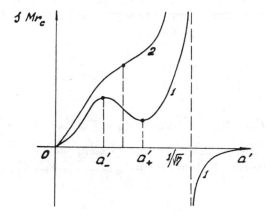

Figure 3.6. Dependence of threshold sMr_c number on the ratio of layers thickness a'.

$$\int_{-\infty}^{\infty} h(x)dx = 0,$$

therefore the ratio of layers thicknesses a' in any region of x change will be larger than a, and in any other region will be smaller than a. If $\mathcal{K}_1(a) \neq 0$, then in one of these regions the threshold value of Marangoni number becomes lower than sMr_*. In this case the conditions for the development of subcritical finite amplitude instability are created. If $\mathcal{K}_1(a) = 0$ and $\mathcal{K}_2(a) < 0$, the finite-amplitude instability arises both in the regions $h > 0$ and $h < 0$. Only as $\mathcal{K}_1(a) = 0$ and $\mathcal{K}_2(a) > 0(a = a'_+)$, we can expect a soft emergence of a supercritical convection.

The above mentioned qualitative considerations are confirmed by the quantitative theory which will be discussed later.

Let us consider that $|Mr - Mr(0)| \sim \varepsilon \ll 1$, parameters W and $Ga\delta$ are of unit order. In addition the disturbances with the wave vectors $|k| \sim \varepsilon^{1/2}$ are of interest, therefore it is worth passing on to "slow" variables $\bar{x} = \varepsilon^{1/2}x, \bar{y} = \varepsilon^{1/2}y$. As follows from the expression for the disturbances decrement, the temporal scale of a longwave disturbances evolution in the vicinity of the threshold Marangoni number is $\bar{t} = \varepsilon^2 t$. Nontrivial equations, describing the longwave thermocapillary convection, are obtained under assumption that the deflection of interface $z = h(x, y)$ from the steady state $z = 0$ is of ε order. Besides the change of variables is performed:

$$h = \varepsilon\bar{h}, \quad T_1 = -s\frac{z-1}{1 + \kappa a} + \varepsilon\theta_1, \quad T_2 = -s\frac{\kappa z - a}{1 + \kappa a} + \varepsilon\theta_2,$$

$$\tag{3.40}$$

$$P_m = \varepsilon P_m, \quad u_m = \varepsilon^{3/2}U_m, \quad v_m = \varepsilon^{3/2}V_m, \quad w_m = \varepsilon^2 W_m$$

(u, v, w are $x-, y-$ and z — velocity components).

Let us expand the functions $\bar{h}, \theta_m, P_m, U_m, V_m, W_m$ in ε:

$$\bar{h} = h^{(0)} + \varepsilon h^{(1)} + \cdots, \tag{3.41}$$

and in the same way for all other variables. We choose

$$Mr = Mr^{(0)} + \varepsilon Mr^{(1)}, \tag{3.42}$$

$$\frac{\partial}{\partial x} = \varepsilon^{1/2}\frac{\partial}{\partial \bar{x}}, \quad \frac{\partial}{\partial y} = \varepsilon^{1/2}\frac{\partial}{\partial \bar{y}}, \quad \frac{\partial}{\partial t} = \varepsilon^2\frac{\partial}{\partial \bar{t}} + \varepsilon^3\frac{\partial}{\partial \bar{t}_1} + \cdots \tag{3.43}$$

Let us substitute (3.40)–(3.43) into complete nonlinear convection equations (1.11)–(1.20) (as $G = 0$) and equate the terms of the same order in ε.

In the lowest order the problem turns out to be solved at $Mr^{(0)} = Mr(0)$ (see formula (2.75)). The solution has the form:

$$\theta_1^{(0)} = h^{(0)} \frac{s(\kappa - 1)}{(1 + \kappa a)^2}(z - 1), \quad P_1^{(0)} = h^{(0)} \frac{Ga\delta\eta a^2}{(1 - \eta a^2)},$$

$$U_1^{(0)} = \frac{\partial h^{(0)}}{\partial \bar{x}} \frac{Ga\delta\eta a^2}{(1 - \eta a^2)} \left[\frac{(z - 1)^2}{2} + \frac{z - 1}{3}\right],$$

$$W_1^{(0)} = \bar{\Delta}_2 h^{(0)} \frac{Ga\delta\eta a^2}{(1 - \eta a^2)} \left[-\frac{(z - 1)^3}{6} - \frac{(z - 1)^2}{6}\right],$$

$$\theta_2^{(0)} = h^{(0)} \frac{s\kappa(\kappa - 1)}{(1 + \kappa a)^2}(z + a),$$

$$P_2^{(0)} = h^{(0)} \frac{Ga\delta}{(1 - \eta a^2)},$$

$$U_2^{(0)} = \frac{\partial h^{(0)}}{\partial \bar{x}} \frac{\eta Ga\delta}{(1 - \eta a^2)} \left[\frac{(z + a)^2}{2} - \frac{a(z + a)}{3}\right],$$

$$W_2^{(0)} = \bar{\Delta}_2 h^{(0)} \frac{\eta Ga\delta}{(1 - \eta a^2)} \left[-\frac{(z + a)^3}{6} + \frac{a(z + a)^2}{6}\right],$$

$$\delta = \rho^{-1} - 1; \quad \bar{\Delta}_2 = \frac{\partial^2}{\partial \bar{x}^2} + \frac{\partial^2}{\partial \bar{y}^2}.$$

The expression for $V_m^{(0)}$ can be obtained from the formulas for $U_m^{(0)}$ by changing $\partial/\partial\bar{x}$ into $\partial/\partial\bar{y}$. In this order function $h^{(0)}(\bar{x}, \bar{y}, \bar{t})$ remains arbitrary.

The solvability condition of the second order equations leads to the evolution equation for function $h^{(0)}$ which can be written in the form:

$$\frac{\partial h^{(0)}}{\partial \bar{t}} = F\bar{\Delta}_2^2 h^{(0)} + BMr^{(1)}\bar{\Delta}_2 h^{(0)} + C\bar{\Delta}_2(h^{(0)})^2, \qquad (3.44)$$

where

$$B = -\frac{s(1 - \eta a^2)\kappa(1 + a)a^2}{2(1 + \kappa a)^2[(1 - \eta a^2)^2 + 4\eta a(1 + a)^2]},$$

$$F = BNMr^{(0)}, \tag{3.45}$$

$$C = -\frac{1}{2}BK_1,$$

K_1 is defined by (3.38). Let us remind that parameter N, defined by formula (2.77), describes the form of a neutral curve in a longwave region:

$$Mr(k) = Mr(0)(1 + k^2N).$$

In equation (3.44) parameter B is always negative, because s and $1 - \eta a^2$ have the same signs due to (2.75). Parameters F and N have opposite signs. If $N < 0$ the disturbances with small k are not most dangerous, and there are no reasons for the emergence of a longwave convection. Therefore, later on we assume that $F < 0$, $N > 0$. The sign of coefficient C coincides with that of K_1, discussed earlier. Value $Mr^{(1)} = (Mr - Mr^{(0)})/\varepsilon$ is positive in a supercritical region and is negative in a subcritical region.

Let $C \neq 0$. Let us introduce new variables:

$$H = \frac{C}{B}h^{(0)}, \quad \tau = -\frac{B^2}{F}\bar{t},$$

$$X = \left(\frac{B}{F}\right)^{1/2}\bar{x}, \quad Y = \left(\frac{B}{F}\right)^{1/2}\bar{y}.$$

Equation (3.44) can be transformed into

$$\frac{\partial H}{\partial \tau} + \Delta_2^2 H + Mr^{(1)}\Delta_2 H + \Delta_2(H^2) = 0, \tag{3.46}$$

where $\Delta_2 = \partial^2/\partial X^2 + \partial^2/\partial Y^2$.

The equation of type (3.46) (in one-dimensional variant) was earlier presented in [134]. In this work longwave convective motions in the anomalous thermocapillary effect conditions have been studied i.e. the case when the dependence of surface tension on temperature is described by

$$\sigma = \sigma_0 + \alpha(T - T_0)^2.$$

The temperature on the interface at the reference state was supposed to be equal to T_0. As far as in the limits of the linear theory for the anomalous thermocapillary effect the steady state is stable, in [134] the coefficient at $\partial^2 H/\partial X^2$ turned out to be negative. As follows from (3.44), the quadratic character of nonlinearity in the equations for the deformation of the surface is not specific for the description of the anomalous thermocapillary effect, but is characteristic of a "usual" thermocapillary convection. We shall consider the steady flow regimes, described by equation (3.46). For one-dimensional motions ($H = H(X)$) the equation (3.46) gets the form:

$$\frac{d^2}{dX^2}\left(\frac{d^2 H}{dX^2} + Mr^{(1)}H + H^2\right) = 0. \qquad (3.47)$$

Since variable H is proportional to the relief height deflection from the middle position, the relation

$$\int_{-\infty}^{\infty} H\,dX = 0 \qquad (3.48)$$

is fulfilled.

Problem (3.47), (3.48) possesses a class of periodic solutions

$$H(X) = -\frac{Mr^{(1)}}{2} - \frac{2-q^2}{3q^2}A + \frac{1-q^2}{q^2}A\frac{1}{dn^2\xi}, \qquad (3.49)$$

where

$$A = \max_x H - \min_x H = \frac{3q^2 Mr^{(1)}}{2[q^2 - 2 + 3E(q)\mathcal{K}^{-1}(q)]} > 0, \qquad (3.50)$$

$$\xi = \frac{\sqrt{A}}{3q}(X - X_0),$$

$dn\xi$ is an elliptical Jacobi function with modulus q; $E(q)$ and $\mathcal{K}(q)$ are complete elliptical integrals, X_0 is an arbitrary constant.

The space period of function $H(X)$ is equal to

$$L = \frac{2\pi}{k}, \quad k = \frac{\pi\sqrt{A/6}}{qK(q)}. \tag{3.51}$$

Let us consider first the region $Mr^{(1)} > 0$, (i.e. $Mr > Mr^{(0)}$). In this region the condition $A > 0$ leads to the relation $q^2 - 2 + 3E(q)K^{-1}(q) > 0$ which is fulfilled at $0 < q < q_*$, where $q \simeq 0.98038$. The dependence of the relief amplitude A on the wave number k is given by parametric formulas (3.49), (3.50) and is shown in Fig. 3.7 (line 1). One can see that solutions bifurcation takes place in the stability region of the reference state $k > k_*$, where $k_* = [Mr^{(1)}]^{1/2}$ is the neutral disturbances wave number. In accordance with the known bifurcation theory results [36], such solutions are unstable. In the functional space the saddle point corresponds to them. Its stable manifold separates the region of growing and decaying finite-amplitude disturbances imposed on the steady state. In such a way variable A determines the amplitude, lower which the disturbances with the given wave number decrease and when higher — increase. The dependence of parameter q on k is shown in Fig. 3.8 (line 1). Note that even at a small deflection of k from unit parameter q is close to q_*, so the relief considerably differs from the sinusoidal one.

As $Mr^{(1)} = 0$ ($Mr = Mr^{(0)}$) solution (3.49) satisfies correlation (3.48) as $q = q_*$ irrespective of value A. The dependence of the amplitude on the wave number has the form

$$A = 6\left(\frac{q_* K(q_*)}{\pi}\right)^2 k^2$$

for all k (see Fig. 3.7; line 2).

In a subcritical region $Mr^{(1)} < 0$ ($Mr < Mr^{(0)}$) we have $q_* < q < 1$.

Solution (3.49) exists for any value of the wave number. The dependence of amplitude $|A|$ and parameter q on wave number k is shown in Fig. 3.7 (line 3) and 3.8 (line 3). The minimum value of A is reached at $k \to 0$ ($q \to 1$) for solution

$$H = \frac{3}{2}|Mr^{(1)}|ch^{-2}\frac{|Mr^{(1)}|^{1/2}X}{2}.$$

As k increases amplitude A, determining the stability region boundary with respect to the finite disturbances, rapidly increases, and variable q staying close to unity, the form of a neutral finite-amplitude disturbance differs greatly from the sinusoidal one. Apparently, this circumstance is explained by the fact, that in numerical experiments of V. V. Puch-

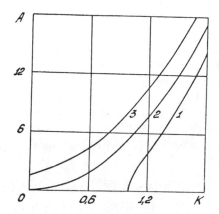

Figure 3.7. Dependence of the relief amplitude on the wave number for values $Mr^{(2)}=1$ (line 1); 0 (2); -1 (3).

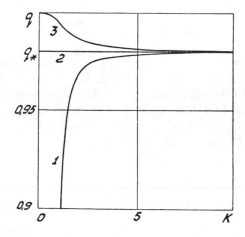

Figure 3.8. Dependence of parameter q on the wave number for values $Mr^{(2)}=1$ (line 1); 0 (2); -1 (3).

nachev [135] the finite-amplitude instability of the reference state has not been observed with space-periodic disturbances, being imposed. But at the same time for the solitary wave form disturbances such instability has been observed.

Equation (3.46) also possesses two-dimensional space-periodic solutions with a different space structure (square, rectangular, hexagonal cells). There is no chance of getting such solutions analytically for an arbitrary wave number. Let us restrict the consideration by the bifurcation analysis of two-dimensional solutions near stability region boundary $k = k_* = (Mr^{(1)})^{(1/2)}$ for $Mr^{(1)} > 0$. The solution in the form of rolls (3.49) at small $k - k_*$ can be written in the form

$$H = 2[3k_*(k - k_*)]^{1/2} \cos kX + \mathcal{O}(k - k_*). \tag{3.52}$$

For rectangular cells the expansion of (3.46) equation solution into the series in the amplitude gives the following expression:

$$H = \left[a(s^2)(k - k_*)k_*\right]^{1/2} \cos k\sqrt{1 - s^2}X \cos skY + \mathcal{O}(k - k_*),$$

where

$$a(s^2) = \frac{48(3 - 4s^2)(1 - 4s^2)}{16s^4 - 16s^2 - 9} \quad \left(s \neq \frac{1}{2}\right).$$

Here $s = \sin(\varphi/2)$; φ is the angle between the basis wave vectors of the rectangular structure. As $\varphi < 60°(s < 1/2)$ $a(s^2) < 0$, which corresponds to the soft bifurcation of the structure in the region $k < k_*$. If $\varphi > 60°(s > 1/2)$, then $a(s^2) > 0$ is a hard bifurcation, in particular this is the character of bifurcation for square cells ($\varphi = 90°, s = \sqrt{2}/2$). Value $\varphi = 60°$ corresponds to the hexagonal structure, for which we obtain the two-hand side bifurcation

$$H = 2(k - k_*)(\cos kX + 2\cos\frac{kX}{2}\cos\frac{\sqrt{3}}{2}kY) + \mathcal{O}(k - k_*). \tag{3.53}$$

As observed earlier, the hard-bifurcation steady solution defines the stability threshold of the steady state with respect to the finite-amplitude disturbances. The comparison of expressions (3.52) and (3.53) shows that the amplitude of disturbance with $k > k_*$ in the hexagon form, leading to the instability, is considerably lower, than the amplitude of disturbance in the form of rolls.

Thus, the longwave thermocapillary instability in a two-layer system has a subcritical ("hard") character. Nonlinear effects intensify the thermocapillary instability and do not lead to the establishment of the stable stationary relief (in the region of equation (3.44) employment).

We consider now the case, when constant C in (3.44) vanishes. As has been mentioned above, this can take place at some values of $a = a'_{\pm}$, if parameters η and κ are located within indicated regions. In order to get the closed nonlinear equation, defining the evolution of interface deflection h, instead of (3.40) it is necessary to impose:

$$h = \varepsilon^{1/2}\bar{h}, \quad T_1 = -s\frac{z-1}{1+\kappa a} + \varepsilon^{1/2}\theta_1,$$

$$T_2 = -s\frac{\kappa z - 1}{1+\kappa a} + \varepsilon^{1/2}\theta_2, p_m = \varepsilon^{1/2}P_m, u_m = \varepsilon U_m, \qquad (3.54)$$

$$v_m = \varepsilon V_m, \quad w_m = \varepsilon^{3/2}W_m \quad (m = 1, 2)$$

and instead of (3.41) we put

$$\bar{h} = h^{(0)} + \varepsilon^{1/2}h^{(1/2)} + \varepsilon h^{(1)} + \cdots \qquad (3.55)$$

Then from the solvability condition of the problem in the order of $\varepsilon^{(1/2)}$ we get $C = 0$, and from the solvability condition in the order of ε we obtain the evolution equation for $h^{(0)}$:

$$\frac{\partial h^{(0)}}{\partial \bar{t}} = F\bar{\Delta}_2^2 h^{(0)} + BMr^{(1)}\bar{\Delta}_2 h^{(0)} + D\bar{\Delta}_2(h^{(0)3}), \qquad (3.56)$$

where values F and B are the same as in equation (3.44), $D = -BK_2/6$; coefficient K_2, determined by formula (3.39) may be of a different sign.

We shall restrict the consideration by the case when $K_2 > 0$, for which, as it was noted earlier, one can expect a soft emergence of convection.

Let us introduce new variables:

$$H = \left(-\frac{D}{B}\right)^{1/2} h^{(0)}, \quad \tau = -\frac{B^2}{F}\bar{t}, \qquad (3.57)$$

$$X = \left(\frac{B}{F}\right)^{1/2}\bar{x}, \quad Y = \left(\frac{B}{F}\right)^{1/2}\bar{y}.$$

By so doing (3.56) takes the form of the so-called *Cahn − Hilliard* equation [122]

$$\frac{\partial H}{\partial \tau} + \Delta_2^2 H + Mr^{(1)}\Delta_2 H - \Delta_2(H^3) = 0.$$

Consider the nonlinear disturbance interaction with different wave vectors. We shall impose disturbance on the reference state:

$$H(R, \tau) = \sum_{n=1}^{N} (H_n(\tau) e^{ik_n R} + H_n(\tau) e^{-ik_n R}); \quad R = (X, Y),$$

where $|k_n| = k$ for all n. The amplitudes evolution H_n is described by the system of equations

$$H_n = k^2 \left[\gamma - e|H_n|^2 - f \sum_{m \neq n} |H_m|^2 \right] H_n,$$

$$\gamma = M r^{(1)} - k^2,$$

$$e = 3, \quad f = 6.$$

$$(3.58)$$

As shown in [152], when $f > e$ the only steady form of motion are convective rolls. Therefore later on we shall restrict ourselves by the investigation of the flows with $H = H(X)$. The steady convective structure is described by equation

$$H^{IV} + M r^{(1)} H'' - (H^3)'' = 0 \tag{3.59}$$

(the prime means differentiation with respect to X). The limited solutions of equation (3.59) exist as $M r^{(1)} > 0$ ("soft" instability) and have the form

$$H = \left(\frac{2q^2 M r^{(1)}}{q^2 + 1} \right)^{1/2} sn \left[(X - X_0) \left(\frac{M r^{(1)}}{q^2 + 1} \right)^{1/2} \right], \quad 0 < q < 1, \tag{3.60}$$

where q is a modulus of an elliptical Jacobi function and X_0 is a constant. The period of the function (3.60) is

$$L = \frac{2\pi}{k}, \quad k = \left(\frac{M r^{(1)}}{q^2 + 1} \right)^{1/2} \frac{\pi}{2K(q)}; \tag{3.61}$$

here $K(q)$ is a complete elliptical integral of the second type. As $k \to k_* = (M r^{(1)})^{(1/2)} q \to 0$,

$$H = 2 \left[\frac{2}{3} k_* (k_* - k) \right]^{1/2} \sin k(X - X_0).$$

As $k = 0, q = 1$,

$$H = \pm(Mr)^{1/2}th\left[(X - X_0)\left(\frac{Mr^{(1)}}{2}\right)^{1/2}\right]. \tag{3.62}$$

Solution (3.62) is a steady "step" soliton, supported by a convective roll centred at $X = 0$. At long distances from the step the thicknesses of layers differ from initials a_1 and a_2, that is why Mr for the system becomes lower than the threshold one, defined by formula (3.37).

As a result, the convective instability does not arise far away from the step.

For small values of k solution (3.60) describes a periodic system of solitons, situated at π/k distances from each other. The softly bifurcating set of solutions (3.60) is stable with respect to the disturbances with the same period (3.61) as that of the solution itself. The stability of these solutions with respect to the arbitrary infinitely small disturbances (in application to other physical problems) was investigated in [102]. It was observed that all the solutions, except (3.62), are unstable. It is necessary to notice, however, that the increment of increasing disturbances for the motions with $k/k_* \ll 1$ is exponentially small, so that the destruction process of such a structure can be very long. Now let us discuss soft rolls bifurcation picture realized in the case when $\mathcal{K}_1 = 0, \mathcal{K}_2 > 0$ transforming to a hard subcritical instability picture characteristic of $\mathcal{K}_1 \neq 0$ at continuous change of the system parameters. For this purpose we shall consider a situation when parameter $\mathcal{K}_1$ differs from zero, but it is small: $\mathcal{K}_1 = O(\varepsilon^{1/2})$. Then the evolution equation for function $h^{(0)}$ which is defined by relations (3.54) and (3.55), takes the form:

$$\frac{\partial h^{(0)}}{\partial \bar{t}} = F\bar{\Delta}_2^2 h^{(0)} + BMr^{(1)}\bar{\Delta}_2 h_+^{(0)}\varepsilon^{-1/2}C\bar{\Delta}_2(h^{(0)})^2 + D\bar{\Delta}_2(h^{(0)})^3. \tag{3.63}$$

Let us exchange variables

$$H = \left(-\frac{D}{B}\right)^{1/2}\left[h^{(0)} + \frac{C}{3\varepsilon^{1/2}D}\right], \quad \tau = -\frac{B^2}{E}\bar{t},$$

$$X = \left(\frac{B}{F}\right)^{1/2}\bar{x}, \quad Y = \left(\frac{B}{F}\right)^{1/2}\bar{y}.$$

Then (3.63) reduces to the form

$$\frac{\partial H}{\partial \tau} + \Delta_2^2 H + \widetilde{M}r^{(1)}\Delta_2 H - \Delta_2(H^3) = 0, \tag{3.64}$$

where

$$\widetilde{M}r^{(1)} = Mr^{(1)} + \frac{C^2}{(-3\varepsilon BD)} \tag{3.65}$$

(the last term on the right-hand side of (3.65) is positive). The conservation condition of the middle layers thicknesses in both fluids (3.48) gives

$$\lim_{L\to\infty} \frac{1}{2L} \int_{-L}^{L} H\,dX = \bar{H}, \tag{3.66}$$

instead of (3.48), where

$$\bar{H} = \frac{C}{3(-\varepsilon BD)^{1/2}}.$$

For steady one-dimensional solutions we have:

$$\frac{d^2}{dX^2}\left(\frac{d^2 H}{dX^2} + \widetilde{M}r^{(1)}H - H^3\right) = 0,$$

or

$$\frac{d^2 H}{dX^2} + \widetilde{M}r^{(1)}H - H^3 = C_1 X + C_2, \tag{3.67}$$

where C_1 and C_2 are constants. From the limitation condition at $X \to \pm\infty$ it follows $C_1 = 0$. One can easily see that the problem of getting steady convective structures is formally equivalent to the problem of nonlinear oscillations of the particle with potential energy

$$U(H) = \frac{1}{2}\widetilde{M}r^{(1)}H^2 - \frac{1}{4}H^4 - C_2 H$$

(coordinate X plays the role of time). The limited solutions with $\bar{H} \neq 0$ exist in the region of the parameter values

$$\widetilde{M}r^{(1)} > 0, \quad 0 < C_2 < 2(\widetilde{M}r^{(1)}/3)^{3/2},$$

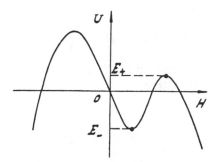

Figure 3.9. "Effective potential".

(in which "effective potential" $U(H)$ has the form given in Fig. 3.9) and can be expressed by elliptical functions [122]. It is evident that for all fixed $Mr^{(1)}$ and D the problem possesses a set of periodical solutions, for which variable

$$E = \frac{1}{2}\left(\frac{dH}{dX}\right)^2 + U(H)$$

is situated in the interval $E_- < E < E_+$, and an aperiodical solution with $E = E_+$. Note that all the solutions with $\overline{H} \neq 0$, in contrast to (3.62) are nonmonotonous. All the enumerated structures turn to be unstable. This conclusion follows from the result of the article [122], according to which steady solution stability (3.64) can be stated on the basis of analysis of the auxiliary problem spectrum of eigenvalues

$$\Delta_2\varphi + (\widetilde{M}r^{(1)} - 3H^2)\varphi = \sigma\varphi; \quad \lim_{x\to\pm\infty} |\varphi| < \infty \qquad (3.68)$$

($\sigma \leq 0$ corresponds to the case of stability). As far as the problem is always of a nonconstant sign solution

$$\varphi = \frac{dH}{dX}, \quad \sigma = 0,$$

and maximum value $\sigma = \sigma_{max}$ is not degenerate and corresponds to a sign-constant eigenfunction, it leads to $\sigma_{max} > 0$. Under real conditions, however, the layers always have a finite size. It is possible to show that the presence of a solid heat-insulated lateral walls leads to boundary conditions

$$X = \pm L: \quad H = \frac{dH}{dX} = 0. \tag{3.69}$$

At $L \gg 1$ and $\overline{H} < 1$ problem (3.67), (3.69) possesses two monotonous solutions, satisfying condition

$$\frac{1}{2L} \int_{-L}^{L} H \, dX = \overline{H}$$

which can be written with an exponential accuracy in the form (3.62), where $X_0 = \mp L\overline{H}$. The same as in case $\overline{H} = 0$, these solutions are stable. The region of their existence at $L \gg 1$ has the form $Mr^{(1)} > 0$, i.e. $Mr^{(1)} > -C^2/(-3\varepsilon BD)$.

Thus, for any small $C \neq 0$ a subcritical instability of the steady state takes place with respect to the steady relief generation in a "step"-form.

In case $C \gg \varepsilon^{1/2}$, the calculation of a stationary relief and the depth of subcritisity lies beyond the capacity of a small-parameter method.

CHAPTER 4

Finite-Amplitude Thermogravitational Convection in Two-Layer Systems

The investigation of a well-developed convection far from the threshold of stability may be based on numerical methods only. In this chapter different variants of the finite-differences method are used for the numerical solution of thermogravitational convection complete equations. Thermocapillary effects are not taken into consideration.

The finite-differences schemes used for the approximation of nonlinear convective equations and the procedure of obtaining solutions are described in Section 4.1. Sections 4.2–4.4 are devoted to the investigation of the convective motions, arising as a result of the mechanical reference state instability of two-layer systems.

In Section 4.2 steady convective motions, arising when the equilibrium temperature gradient is directed vertically down (heating from below) are studied. The attention is paid to the investigation of the supercritical motions form and their stability. The oscillatory convective motions differing in a space structure and in the character of its change in time are studied in Section 4.3. In Section 4.4 the nonlinear convection when heating from above is considered. The calculation results of convective flows as the temerature gradient is directed along the interface are described in Section 4.5.

4.1.   Method of finite differences

Let us consider a system of two different viscous immiscible fluids filling a rectangular cylinder $0 \le x \le l, -a_2 < z < a_1, -\infty < y < \infty$; the interface between the fluids is assumed to be horizontal and undeformable: $z = 0$ (see Fig. 2.39). As in 1 Section 2.8, the horizontal cavity boundaries are chosen to be solid and isothermal and on the solid vertical walls the temperature distribution, corresponding to heating from below and providing the mechanical steady state opportunity is given (see (2.78), (2.79)).

Introducing dimensionless variables in the same way as in Section 1.2, we get the following boundary problem (see (1.11)–(1.14), (1.16), (1.17), (1.19), (1.20), (1.23)):

$$\frac{\partial v_1}{\partial t} + (v_1 \nabla) v_1 = -\nabla p_1 \Delta v_1 + GT_1 \gamma,$$

$$\frac{\partial T_1}{\partial t} + v_1 \nabla T_1 = \frac{1}{P} \Delta T_1, \quad \text{div} v_1 = 0,$$

$$\frac{\partial v_2}{\partial t} + (v_2 \nabla) v_2 = -\rho \nabla p_2 + \frac{1}{\nu} \Delta v_2 + \frac{G}{\beta} T_2 \gamma,$$

$$\frac{\partial T_2}{\partial t} + v_2 \nabla T_2 = \frac{1}{\chi P} \Delta T_2, \quad \text{div} v_2 = 0,$$

$$z = 1 : v_1 = 0, \quad T_1 = 0,$$

$$z = -a : v_2 = 0, \quad T_2 = 1,$$

$$z = 0 : \eta \frac{\partial v_{1,x}}{\partial z} - \frac{\partial v_{2,x}}{\partial z} = 0, \quad \eta \frac{\partial v_{1,y}}{\partial z} - \frac{\partial v_{2,y}}{\partial z} = 0,$$

$$v_{1,x} = v_{2,x}, \quad v_{1,y} = v_{2,y}, \quad v_{1,z} = v_{2,z} = 0,$$

$$T_1 = T_2, \quad \kappa \frac{\partial T_1}{\partial z} = \frac{\partial T_2}{\partial z};$$

$$x = 0, L; \quad z \geq 0 : v_1 = 0, \quad T_1 = \frac{1-z}{1+\kappa a};$$

$$x = 0, L; \quad z \leq 0 : v_2 = 0, \quad T_2 = \frac{1-\kappa z}{1+\kappa a};$$

$$L = \frac{l}{a_1}.$$

As in 1 Section 2.8 we shall restrict ourselves with the consideration of homogeneous with respect to the y — coordinate two-dimensional motions ($v_{1,y} = v_{2,y} = 0$), for the description of which it is possible to introduce the stream function:

$$v_{m,x} = \frac{\partial \psi_m}{\partial z}, \quad v_{m,z} = -\frac{\partial \psi_m}{\partial x} \quad (m = 1, 2).$$

In writing the equations we need one more variable, the velocity vortex

$$\varphi_m = \frac{\partial v_{m,z}}{\partial x} - \frac{\partial v_{m,x}}{\partial z},$$

connected with ψ_m by

$$\Delta \psi_m = -\varphi_m, \quad \Delta = \frac{\partial^2}{\partial x^2} + \frac{\partial^2}{\partial z^2}.$$

In new variables the equations and conditions on the interface take the form:

$$\frac{\partial \varphi_1}{\partial t} + \left(\frac{\partial \psi_1}{\partial z} \frac{\partial \varphi_1}{\partial x} - \frac{\partial \psi_1}{\partial x} \frac{\partial \varphi_1}{\partial z} \right) = \Delta \varphi_1 + G \frac{\partial T_1}{\partial x}, \qquad (4.1)$$

$$\frac{\partial T_1}{\partial t} + \left(\frac{\partial \psi_1}{\partial z} \frac{\partial T_1}{\partial x} - \frac{\partial \psi_1}{\partial x} \frac{\partial T_1}{\partial z} \right) = \frac{1}{P} \Delta T_1, \qquad (4.2)$$

$$\Delta \psi_1 = -\varphi_1, \qquad (4.3)$$

$$\frac{\partial \varphi_2}{\partial t} + \left(\frac{\partial \psi_2}{\partial z} \frac{\partial \varphi_2}{\partial x} - \frac{\partial \psi_2}{\partial x} \frac{\partial \varphi_2}{\partial z} \right) = \frac{1}{\nu} \Delta \varphi_2 + \frac{G}{\beta} \frac{\partial T_2}{\partial x}, \qquad (4.4)$$

$$\frac{\partial T_2}{\partial t} + \left(\frac{\partial \psi_2}{\partial z} \frac{\partial T_2}{\partial x} - \frac{\partial \psi_2}{\partial x} \frac{\partial T_2}{\partial z} \right) = \frac{1}{\chi P} \Delta T_2, \qquad (4.5)$$

$$\Delta \psi_2 = -\varphi_2, \qquad (4.6)$$

$$z = 0 : \psi_1 = \psi_2 = 0, \quad \frac{\partial \psi_1}{\partial z} = \frac{\partial \psi_2}{\partial z}, \quad \eta \varphi_1 = \varphi_2, \qquad (4.7)$$

$$T_1 = T_2, \quad \kappa \frac{\partial T_1}{\partial z} = \frac{\partial T_2}{\partial z}, \qquad (4.8)$$

Conditions on the cavity boundaries in new variables will be rewritten in the following way: on solid horizontal boundaries

$$z = 1 : \psi_1 = \frac{\partial \psi_1}{\partial z} = 0, \quad T_1 = 0,$$

$$z = -a : \psi_2 = \frac{\partial \psi_2}{\partial z} = 0, \quad T_2 = 1; \tag{4.9}$$

on the lateral boundaries with solid walls

$$x = 0, L : \psi_1 = \frac{\partial \psi_1}{\partial x} = 0, \quad T_1 = \frac{1-z}{1+\kappa a}, \quad (z \geq 0),$$

$$x = 0, L : \psi_2 = \frac{\partial \psi_2}{\partial x} = 0, \quad T_2 = \frac{1-\kappa z}{1+\kappa a}, \quad (z \leq 0). \tag{4.10}$$

In equations (4.1)–(4.8) and boundary conditions (4.9), (4.10) seven physical parameters $(G, P, \eta, \nu, \kappa, \chi, \beta)$ and two geometrical parameters (L, a) are used. Observe that the linearized problem for the case of a monotonous instability is defined by only seven independent parameters $(R = GP, \eta, \zeta = \nu/\beta, \kappa, \chi, L, a)$.

The boundary value problem (4.1)–(4.10) was solved by the method of finite differences; two schemes of the second order accuracy at the space mesh step — explicit scheme with a centrally-differential approximation of space derivatives and the scheme of a linear-tranverse drive have been used in parallel [178, 191]. We shall divide the calculated region into two subregions — $1: 0 \leq x \leq L, 0 \leq z \leq 1; 2: 0 \leq x \leq L, -a \leq z \leq 0$. In each subregion we shall introduce a uniform mesh:

$$1: \quad x_i = i\Delta x; \quad i = 0, 1, \ldots, I; \quad \Delta x = L/I;$$

$$z_j = j\Delta z_1; \quad j = 0, 1, \ldots, J_1; \Delta z_1 = 1/J_1;$$

$$2: \quad x_i = i\Delta x; \quad i = 0, 1, \ldots, I; \quad \Delta x = L/I;$$

$$z_j = j\Delta z_2; \quad j = 0, 1, \ldots, J_2; \quad \Delta z_2 = a/J_2.$$

We shall notate the time step as Δt. Variables ψ_m, φ_m and T_m are defined in the nodes at the intersections of vertical (i-th) and horizontal (j-th) mesh lines at the n-th time layer.

In case of an explicit scheme application the finite- different analogue of equations (4.1)–(4.8) was written in the form:

$$\varphi_{i,j}^{n+1} = \varphi_{i,j}^{n} + \Delta t \left\{ c_m \Delta \varphi_{i,j}^{n} + \frac{Gb_m}{2\Delta x} \left(T_{i+1,j}^{n} - T_{i-1,j}^{n} \right) - \frac{1}{4\Delta x \Delta z_m} \times \right.$$

$$\left. \left[\left(\psi_{i,j+1}^{n} - \psi_{i,j-1}^{n} \right) \left(\varphi_{i+1,j}^{n} - \varphi_{i-1,j}^{n} \right) - \left(\psi_{i+1,j}^{n} - \psi_{i-1,j}^{n} \right) \left(\varphi_{i,j+1}^{n} - \varphi_{i,j-1}^{n} \right) \right] \right\}$$

$$\tag{4.11}$$

$$\varphi_{i,j}^{n+1} = -\Delta \psi_{i,j}^{n+1}, \tag{4.12}$$

$$T_{i,j}^{n+1} = T_{i,j}^{n} + \Delta t \left\{ \frac{d_m}{P} \Delta T_{i,j}^{n} - \frac{1}{4\Delta x \Delta z_m} \left[\left(\psi_{i,j+1}^{n+1} - \psi_{i,j-1}^{n+1} \right) \times \right. \right.$$

$$\left. \left. \left(T_{i+1,j}^{n} - T_{i-1,j}^{n} \right) - \left(\psi_{i+1,j}^{n+1} - \psi_{i-1,j}^{n+1} \right) \left(T_{i,j+1}^{n} - T_{i,j-1}^{n} \right) \right] \right\}. \tag{4.13}$$

In the notations $(\varphi_m)_{i,j}, (\psi_m)_{i,j}$ and$(T_m)_{i,j}$ index $m = 1, 2$ is omitted; $b_1 = c_1 = d_1 = 1, b_2 = 1/\beta, c_2 = 1/\nu, d_2 = 1/\chi$.

As φ^{n+1} and T^{n+1} are calculated by the variables values on the n-th step directly by formulas (4.11), (4.13), the definition of the ψ^{n+1} field values requires Poisson equation (4.12) iteration at each time step. The iteration was done by the Liebman method with a consequent upper relaxation [178]:

$$\psi_{i,j}^{n,s+1} = \psi_{i,j}^{n,s} + \omega \left\{ \frac{1}{2[(\Delta x)^2 + (\Delta z_m)^2]} \left[(\psi_{i+1,j}^{n,s} + \psi_{i-1,j}^{n,s+1})(\Delta z_m)^2 + \right. \right.$$

$$\left. \left. \left(\psi_{i,j+1}^{n,s} + \psi_{i,j-1}^{n,s+1} \right) (\Delta x)^2 + (\Delta x)^2 (\Delta z_m)^2 \varphi_{i,j}^{n,s} \right] - \psi_{i,j}^{n,s} \right\}. \tag{4.14}$$

Here s is the iteration number and ω is the relaxation parameter. In accordance with the theory of method $1 < \omega < 2$; the optimal value of relaxation parameter ω_m for region m ($m = 1, 2$) can be obtained by formula:

$$\omega_m = \frac{2}{1 - \lambda_m^2},$$

$$\lambda_m = \frac{(\Delta z_m)^2 \cos(\pi \Delta x) + (\Delta x)^2 \cos(\pi \Delta z_m)}{(\Delta x)^2 + (\Delta z_m)^2}.$$

When the scheme of a linear-tranverse drive is used [191] which is one of the variants of the fraction step method, the transition to the variables at the next time layer (time step) is done in two stages: from the n-th layer on the $(n + 1/2)$-th auxiliary layer and then from $(n + 1/2)$-th on the $(n + 1)$-th layer. In the scheme of a linear-tranverse drive at the first fracton step space derivatives by x are approximated implicitly and derivatives by z are approximated explicity, and at the second time step process is reversed. Let us write down, for example, the corresponding approximations of the equation for the vortex at the first fraction step.

$$\frac{\varphi_{i,j}^{n+1/2} - \varphi_{i,j}^{n}}{\Delta t/2} =$$

$$-\left[\left(\frac{\psi_{i,j+1}^{n} - \psi_{i,j-1}^{n}}{2\Delta z_m}\right)\left(\frac{\varphi_{i+1,j}^{n+1/2} - \varphi_{i-1,j}^{n+1/2}}{2\Delta x}\right) - \right.$$

$$\left. \left(\frac{\psi_{i+1,j}^{n} - \psi_{i-1,j}^{n}}{2\Delta x}\right)\left(\frac{\varphi_{i,j+1}^{n} - \varphi_{i,j-1}^{n}}{2\Delta z_m}\right)\right] +$$

$$c_m\left[\left(\frac{\varphi_{i+1,j}^{n+1/2} - 2\varphi_{i,j}^{n+1/2} + \varphi_{i-1,j}^{n+1/2}}{(\Delta x)^2}\right) + \right.$$

$$\left. \left(\frac{\varphi_{i,j+1}^{n} - 2\varphi_{i,j}^{n} + \varphi_{i,j-1}^{n}}{(\Delta z_m)^2}\right)\right] +$$

$$Gb_m\left(\frac{T_{i+1,j}^{n+1} - T_{i-1,j}^{n+1}}{2\Delta x}\right);$$

at the second fraction step:

$$\frac{\varphi_{i,j}^{n+1} - \varphi_{i,j}^{n+1/2}}{\Delta t/2} =$$

$$-\left[\left(\frac{\psi_{i,j+1}^{n} - \psi_{i,j-1}^{n}}{2\Delta z_m}\right)\left(\frac{\varphi_{i+1,j}^{n+1/2} - \varphi_{i-1,j}^{n+1/2}}{2\Delta x}\right) - \right.$$

$$\left.\left(\frac{\psi_{i+1,j}^{n} - \psi_{i-1,j}^{n}}{2\Delta x}\right)\left(\frac{\varphi_{i,j+1}^{n+1} - \varphi_{i,j-1}^{n+1}}{2\Delta z_m}\right)\right]+$$

$$c_m\left[\left(\frac{\varphi_{i+1,j}^{n+1/2} - 2\varphi_{i,j}^{n+1/2} + \varphi_{i-1,j}^{n+1/2}}{(\Delta x)^2}\right) + \right.$$

$$\left.\left(\frac{\varphi_{i,j+1}^{n+1} - 2\varphi_{i,j}^{n+1} + \varphi_{i,j-1}^{n+1}}{(\Delta z_m)^2}\right)\right]+$$

$$Gb_m\left(\frac{T_{i+1,j}^{n+1} - T_{i-1,j}^{n+1}}{2\Delta x}\right).$$

The Poisson equation for the field values φ determinations the same way as in case of an explicit scheme is iterated by formula (4.14). The results obtained with different schemes turn out to be practicaly the same.

To complete the description of solution procedure it is necessary to write the formulas for calculation of the vortex values on the boundaries of the region. The values of φ on the solid boundaries are calculated by Thom formula [180]. For example, on the boundary parallel to the z-axis $(x = 0)$ we have :

$$\varphi_m(0, z) = -\frac{2\psi_m(\Delta x, z)}{(\Delta x)^2} \quad (m = 1, 2).$$

To calculate φ and T on the interface the correlations following from conditions (4.7), (4.8) have been used

$$\varphi_1(x, 0) = \frac{-2[\psi_1(x, \Delta z_1)\Delta z_2 + \psi_2(x, -\Delta z_2)\Delta z_1]}{\Delta z_1 \Delta z_2 (\Delta z_1 + \eta\Delta z_2)},$$

$$\varphi_2(x,0) = \eta\varphi_1(x,0),$$

$$T_1(x,0) = T_2(x,0) = \frac{T_1(x,\Delta z_1)\kappa\Delta z_2 + T_2(x,-\Delta z_2)\Delta z_1}{\Delta z_1 + \kappa\Delta z_2}.$$

Besides, for calculating of φ values on the solid boundaries Kuskova-Chudov [90] formulas have been used. For example, on the boundary parallel to z axis we have

$$\varphi_m(0,z) = \frac{\psi_m(2\Delta x, z) - 8\psi_m(\Delta x, z)}{2(\Delta x)^2}$$

The value of time step was chosen in the following way. The minimum space step of the calculated mesh $\Delta h = \min(\Delta x, \Delta z_1, \Delta z_2)$ was defined, the typical times of different diffusion processes were compared

$$\tau_{H_1} = (\Delta h)^2, \quad \tau_{T_1} = (\Delta h)^2 P,$$

$$\tau_{H_2} = (\Delta h)^2 \nu, \quad \tau_{T_2} = (\Delta h)^2 \chi P$$

(4.15)

and the smallest of them was chosen :

$$\tau = \min(\tau_{H_1}, \tau_{T_1}, \tau_{H_2}, \tau_{T_2}).$$

After that the time step was given by

$$\Delta t = \frac{\tau}{2(2 + |\psi_{\max}|)},$$

where $|\psi_{\max}|$ is the maximum value of a stream function modulus in the entire range.

The described method is not, of course, the only possible one for the calculation of convection in the systems with an interface. Finite-difference methods for convection equations system solutions in initial variables $v_m, T_m, p_m (m = 1, 2)$ are often referred too (see Section 4.5). For the calculations of nonlinear convection regimes in two-layer systems the finite-element method can also be used [39, 86, 159].

4.2. Finite-amplitude convection when heating from below

1. Procedure of calculation. In this paragraph the results of numerical investigation of steady convection, arising in a rectangular section cavity with solid walls are discussed [57, 62, 163, 164].

The investigation of convective motions is done on the basis of Cauchy problem solution for non-stationary system (4.1)– (4.10). The initial conditions corresponding to disturbances imposed on the system reference state are given. Numerical solution of a boundary value problem permits to model the process of convection appearance and development. In this case the investigation covers both subcritical situations when the disturbances decrease in time and a supercritical region, where the development of disturbances leads to steady motion emergence.

This approach permits us first of all to define the boundary of stability (the critical Grashof number). In this case different kinds of disturbances approximately corresponding to the character of motions possible in the system are to be sorted out (see 1 Section 2.8). The disturbances usually looked like local changes of the vortex field in one or some points and mostly occurred simultaneously in both fluids. The types of disturbances being used in calculations are schematically shown in Fig. 4.1. Steady motions arising as a result of disturbances development being studied, series of calculations differing in initial disturbances picture have been also carried out. As follows from calculation results, the motion structure can be defined (in some cases) by initial conditions.

As far as motion intensity dependence on the Grashof number is concerned (especially in cases with hysteresis) another method of introducing initial conditions was often used: initial fields ψ and T were taken from solution corresponding to another (close to the given) Grashof number.

The iteration accuracy of the Poisson equation varied within 10^{-4}–10^{-7}. As the stabilization method was used, the criterion of steady regimes being reached is the fall of the flow characteristics relative change to 10^{-5}–10^{-7}. This usually takes several units of dimensionless time.

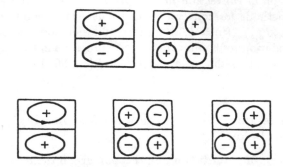

Figure 4.1. The form of initial disturbances.

The number of mesh node depended on the cavity form and changed from 10 × 20 to 48 × 96.

2. Convection in a square cavity. A large number of definitive parameters of the problem does not permit us to get a full picture of fluid regimes and transitions between them. For this reason we shall restrict ourselves with studying separate physical and geometrical parameters' effects on the convective motions structure by means of model systems analysis and by calculations for some concrete media.

In this part we shall fix the geometrical parameters of system ($L = 2; a = 1$) which corresponds to a cavity of square cross section. Typical features of convection emergence and transitions between different motion regimes with the increase in the Grashof number can be illustrated with the simplest model system for which coefficients of dynamic and kinematic viscosities of fluids differ (by a factor of two) ($\eta = \nu = 0.5; \kappa = \chi = \beta = 1$) [164].

When discussing the results of nonlinear calculations it is convinient to use Rayleigh numbers R_1 and R_2 introduced in Section 2.2 and determined by the equilibrium temperature gradients and parameters of the first and the second fluids accordingly:

$$R_1 = \frac{GP}{1 + \kappa a}, \quad R_2 = \frac{GP\kappa}{1 + \kappa a} \frac{\nu \chi a^4}{\beta}, \tag{4.16}$$

which are linked by

$$\frac{R_2}{R_1} = \frac{\kappa \nu \chi a^4}{\beta}. \tag{4.17}$$

Let us recall the main results of the linear theory (*Section* 2.8). With the chosen correlation of parameters, Rayleigh numbers defined by the parameters of upper and lower fluids, according to (4.17) differs by a factor of two: $R_1 = 2R_2$. Because of this, critical conditions are reached with the increase in temperature difference of horizontal boundaries not simultaneously in both fluids. The structure of critical motions, corresponding to lower critical Rayleigh numbers is presented in Fig. 2.40 a, b. Intensive convection arises only in the upper fluid, while in the lower fluid there is a weak flow along the interface induced by the motion in the upper fluid. Since for the given system of parameters $G = 2R_1$, the corresponding critical Grashof numbers are equal to $3.44 \cdot 10^3$ (for the motion with a noneven by x stream function) and $3.59 \cdot 10^3$ (for the motion with an even stream function). Convection in the lower fluid starts at $G \simeq 6.45 \cdot 10^3$.

Let us go on to the calculated results for nonlinear convection regimes. The steady state loses stability at $G > G_* = 3.8 \cdot 10^3$ (i.e. the difference between the threshold Grashof number and the value obtained by the finite-element method is about 10%).

In accordance with linear theory predictions, the intensity of convection in the upper fluid is considerably higher than in the lower one (see Fig. 4.2). The onset of convective motion has a "soft" character. Further increase in the Grashof number leads to the critical conditions in the lower fluid also, and the intensity of convective motion in it grows. The stages of this process are shown in Fig. 4.2–4.4, where the stream lines for different values of the Grashof number are given. Note that convective motion at $G = 4.5 \cdot 10^3$ and $G = 5 \cdot 10^3$ is close in structure to the motion shown in Fig. 2.40 a, corresponding in linear theory to the lower level of spectrum. Further on this form of motion will be called structure a. As $G = 6 \cdot 10^3$ the two lower vortexes, stipulated by Archimedes forces occupy practically all the volume, approaching by intensity to the motion in the upper half of the cavity.

The difference in the character of convection in the upper and the lower fluids is observed also in the amplitude curves. In Fig. 4.5 the dependences of motion intensity on the Grashof number near the threshold in the upper (curve a_1) and the lower (curve a_2) halfs of the region are shown.

Let us consider the changes of supercritical motions structure taking place with further increase in the Grashof number. In Fig. 4.6 the de-

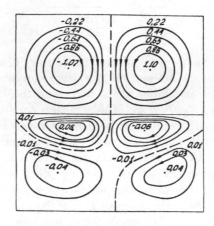

Figure 4.2. Stream lines as $G = 4.5 \cdot 10^3$.

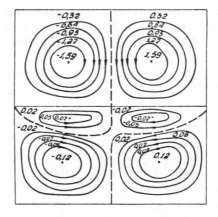

Figure 4.3. Stream lines as $G = 5 \cdot 10^3$.

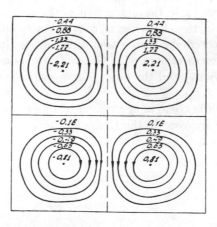

Figure 4.4. Stream lines as $G = 6 \cdot 10^3$.

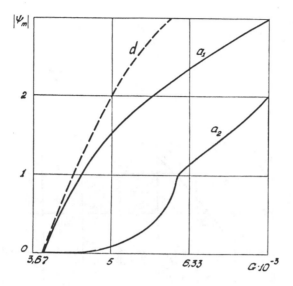

Figure 4.5. The initial section of amplitude curves.

pendence of the stream function maximum value in the upper half of the cavity on the Grashof number at high supercriticities is presented. In region $G < G_1 = 3.8 \cdot 10^4$ the structure a is stable. At high Grashof numbers a four-vortex structure becomes unstable and transforms into a two-vortex structure with counter flows on the interface (sructure b). This transition takes place through an intermediate stage connected with the realization of motion c). In this case hysteresis transitions between the structures a and c are observed.

Thus, as the Grashof number increases, the tendency to horizontal growth of convective structure is observed. Motion d occurs to be metastable — it arises from the disturbance of the appropriate form and becomes practically steady, but after some time is destroyed and turns into motion a.

The initial evolution stages of nonlinear convective motions in the systems with $R_1 \neq R_2$ are apparently universal and weakly depend on concret values of the parameters. With the increase in supercriticity the differences between the motion characteristics for different systems become more obvious. This situation may be illustrated with an example of another model system for which the coefficients of heat conductivity

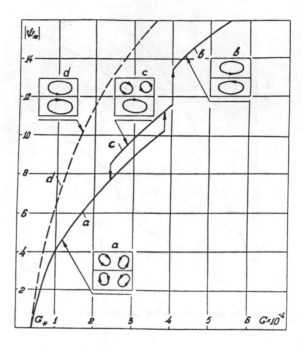

Figure 4.6. Amplitude curves.

and heat diffusivity for the upper fluid are one-half of those for the lower one ($\kappa = \eta = 0.5$; $\eta = \nu = \beta = 1$; $P = 1$; $L = 2$; $a = 1$) [163]. Like in the case $\eta = \nu = 0.5$ discussed above, the Rayleigh number R_1 for the upper half of the cavity turns to be larger than the Rayleigh number R_2 for the lower half, but by a factor of four ($R_1 = 4R_2$). The steady state loses the stability at $G = G_* \simeq 4 \cdot 10^3$; moreover the convection stipulated by Archimedes force develops only in the upper fluid. The motion in the upper half of the cavity has a two-vortex structure, and by the tangential stress on the interface it induces a weak flow in the lower part of the cavity. The motion form slightly above the critical Grashof number completely corresponds to the predictions of the linear theory. The pictures of stream lines showing the evolution of the flow structure are presented in Fig. 4.7 a and 4.8. The respective map of isotherms is shown in Fig. 4.7 b. Note that the values of parameters $\kappa = \chi = 0.5$ have been reported in [1]. Comparing these figures with

Fig. 4.2–4.4, one can see that with the increase in G transition to a well-developed four-vortex motion in both cases proceeds qualitatively in the same manner. However, the observation of the motion intensity change with further increase in the Grashof number reveals essential differences. As $\eta = \nu = 0.5$ $(R_1 = 2R_2)$ the maximum value of function modulus in the lower fluid is always smaller than in the upper fluid (Fig. 4.5). But in case $\kappa = \chi = 0.5$ $(R_1 = 4R_2)$, the motion intensity increases with increase in G in a way shown in Fig. 4.9 where curve 1 reflects the dependence of the maximum of the stream function modulus $|\psi_m(G)|$ for the upper half, and curve 2 does the same for the lower half of the cavity. One can see that in the region of larger subcriticities curve 2 goes higher than curve 1 i.e. convection in the lower fluid (for which the Rayleigh number R_2 is less than R_1) becomes even more intensive than in the upper fluid. This circumstance seems to be connected with the fact that under the conditions of intensive convection mean temperature gradients in both media come close to each other (see Fig. 4.7 *b*) and the Prandtl number for the lower fluid is smaller.

   3. **Convection in a rectangular cavity.** As an example of convection in a rectangular region we shall describe the calculation results of convective regimes for water-dow corning N 200 system (the system parameters are given in Table 1), filling up the horizontally elongated cavity $(L = 6.6; a = 1)$. Calculations were carried out on mesh 48 × 32. The accuracy of the Poisson equation iteration is 10^{-7}.

   At $G < G_* \simeq 200$ the system keeps mechanical steady state. Exceeding of the Grashof critical number is accompanied by the stability loss, and a steady convective motion arises in the system. The relation R_2/R_1 for the system under consideration is equal to 0.0124, therefore convection arises in the upper fluid. The map of isolines corresponding to $G = 350$ is presented in Fig. 4.10. The motion is symmertic with respect to the vertical line $x = L/2$. The vortexes situated near the middle of the cavity possess a largest intensity. In Fig. 4.11 the map of stream lines for $G = 1000$ is shown. The finite-amplitude characteristics are presented in Fig. 4.12. Lines 1–3 correspond to the vortex intensity in the upper fluid and 4–6 — in the lower one (numeration of the vortexes in both layers is carried out from the left to the right). The steady motion keeps until $G \simeq 5000$. Note that as in the case of convection in a square cavity, with increase in the Rayleigh number one can observe the tendency to the enlargement of convective cells.

   4. **Convection in horizontal layers.** Side by side with flows in enclosed cavities a two-dimensional space-periodic convection in horizontal layers was considered. As a calculation region half of a convective cell

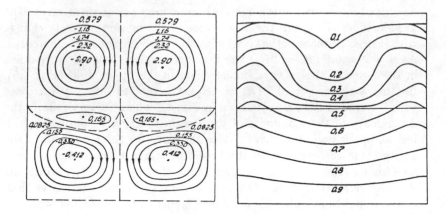

Figure 4.7. Stream lines (a) and isotherms (b) picture as $G = 5 \cdot 10^3$.

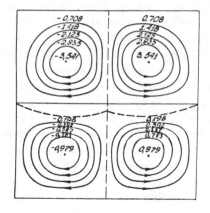

Figure 4.8. Stream lines picture as $G = 6 \cdot 10^3$.

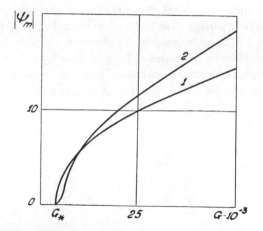

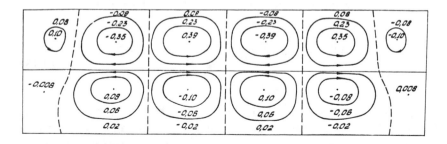

Figure 4.10. Stream lines picture as $G = 350$.

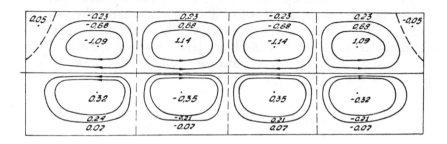

Figure 4.11. Stream lines picture as $G = 1000$.

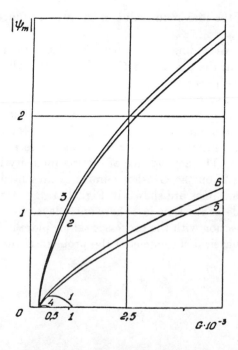

Figure 4.12. Amplitude curves for a water-silicone oil No. 200 system.

$0 \leq x \leq \pi/k, -a \leq z \leq 1$ was chosen (k is dimensionless wave number), on the lateral walls of which the conditions have been put:

$$x = 0, \frac{\pi}{k}: \quad \psi_1 = \varphi_1 = 0, \quad \frac{\partial T_1}{\partial x} = 0 \ (z \geq 0),$$
$$x = 0, \frac{\pi}{k}: \quad \psi_2 = \varphi_2 = 0, \quad \frac{\partial T_2}{\partial x} = 0 \ (z \leq 0).$$
$$(4.18)$$

Let us describe the calculation results of convection for the system of two infinite horizontal layers of immiscible media air and water (see Table 1), restricted by solid plates, on which constant and nonequal temperatures (heating from below) are maintained [57]. The considerable difference between the coefficients of kinematic viscosity and heat diffusivity leading to the difference of typical times between hydrodynamic and heat processes in both media, essentially increases the time necessary for obtaining a steady solution. To accelerate calculations, equations for the upper and the lower layers were solved with different time steps. Namely, the typical time of diffusive processes in both media have been defined:

$$\tau_1 = \min(\tau_{H_1}, \tau_{T_1}), \quad \tau_2 = \min(\tau_{H_2}, \tau_{T_2})$$

and the time step in each layer was chosen according to

$$\Delta t_m = \frac{\tau_m}{2(2 + |\psi_{\max,m}|)}, \quad m = 1, 2,$$

where $|\psi_{\max,m}|$ is the maximum value of the stream function modulus in the m-th layer.

Let $k = \pi/2$. According to the results of numerical calculations, convection arises in the system as $G > G_* = 210$ which is in good agreement with the linear theory.

For the air-water system $R_2/R_1 = 4.66$; therefore above the Grashof number threshold value free-convective motion arises first only in water, while in air the motion induced by the tangencial stress on the interface is realized. Further increase in G leads to the increase of the motion intensity in air too. The dependence of motion intensity in air (line 1) and in water (line 2) on the Grashof number is presented is Fig. 4.13. The maps of stream lines are shown in Fig. 4.14.

Finite-amplitude convection in a horizontal layer has been investigated also in connection with the processes taking place in Earth mantle [32, 39, 86]. Geophysical statement of the problem is characterized by

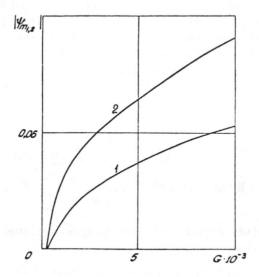

Figure 4.13. Amplitude curves for an air-water system.

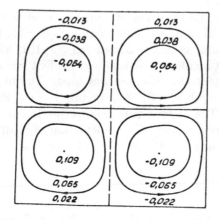

Figure 4.14. Stream lines picture as $G = 1000$.

the infinite value of the Prandtl number (which leads to the disappearance of nonlinear terms in the motion equation); the upper boundary of the first media may be considered to be free. Different variants of conditions for the temperature on the lower boundary are used; investigation of internal heat release is also of interest [39]. To compare geophysical observations the analysis of a convective flow, as a rule, is supplemented by the calculation of relief form of upper boundary $z = a_1 + h_1(x, y)$ on the basis of

$$h_1 = \delta_\beta G^{-1} \left(-p_1 + 2\frac{\partial v_{1,z}}{\partial z} \right)\Big|_{z=a_1},$$

which is deduced by analogy with (1.25).

4.3.   Oscillatory regimes of a thermogravitational convection

In the previous paragraph the calculation results of finite-amplitude steady motions in a two-layer system heated from below have been discussed. Steady motions, however, do not exhaust the whole complex of possible flow regimes.

In this paragraph on the basis of complete nonlinear equations the non-steady convection in the system of two immiscible fluids filling up a cavity of rectangular section is considered [53, 107, 52].

For investigating oscillatory motions, an explicit scheme of stabilization method with central differences was used. The main calculations have been done on mesh 16×32. The accuracy of the Poisson equation iteration is 10^{-7}. The approximation of vortex in solid boundaries is based by Thom formula [180]. The calculation of temperature on the interface was done down to the second order accuracy

$$T_1(x, 0) = T_2(x, O)$$

$$= \frac{[4T_2(x, -\Delta z) - T_2(x, -2\Delta z)] + \kappa[4T_1(x, \Delta z) - T_1(x, 2\Delta z)]}{3(1+\kappa)}$$

$$(\Delta z_1 = \Delta z_2 = \Delta z).$$

$$(4.19)$$

The value of a time step has been chosen as the condition of calculation stability. For nonsteady convective regimes the following verifications were used: 1) decrease of the space mesh step (calculations on mesh 20 × 40); 2) decrease of time step; 3) the use of Kuskova–Chudov formulas [90] for calculation of vortex on the solid walls; 4) improvement of iteration accuracy of the Poisson equation to 10^{-8}.

Transitions between different flow regimes will be considered using the example of water-silicone oil DC 200 at $L = 0.8; a = 1$ (see Table 1). The choice of parameter L is based on the following reasons. The onset of convective oscillations, as a rule, is connected with competition of motions with a different space structure. That is why the simpliest way for the oscillations to arise is the case when the threshold Rayleigh numbers for different modes (one-vortex and two-vortex in the given case) are close. To characterize the flow structure, we introduce variables

$$S_1(t) = \int_0^{L/2} dx \int_0^1 dz \psi_1(x, z, t),$$

$$S_2(t) = \int_{L/2}^L dx \int_0^1 dz \psi_1(x, z, t),$$

$$S_+ = S_1 + S_2, \quad S_- = S_1 - S_2.$$

(4.20)

If $G < G_0 = 1700$ the system keeps the reference state. Above the critical Grashof number the reference state loses stability and a steady convective flow arises (regime 1); moreover, the intensity of motion in the upper fluid is considerably higher than in the lower one. It can be explained by the fact that for this system relation (2.39) of local Rayleigh numbers, defined separately by the parameters of the upper and the lower fluids is $R_2/R_1 = 0.012$. That is why the conditions of convection generation are reached in the upper fluid at considerably smaller values of G than in the lower fluid.

For a steady motion the difference in values S_1 and S_2, stipulated by the difference of boundary conditions on the horizontal boundaries of the upper layer is not large ($|S_1 - S_2|/S_1 < 0.1$); the motion in the upper fluid is mainly one-vortex.

With the increase in G, the steady motion becomes unstable and regular oscillations develop in the system (regime 2). The motion structure in the upper fluid changes considerably: the motion becomes two-vortex, the intensity of vortices periodically changes with time. Value S_- (line 1 in Fig. 4.15; $G = 5700$), characterizing the contribution of a two-vortex motion component is not small and retains its sign during oscillation process; the contribution of one-vortex mode S_+ (line 2 in Fig. 4.15) is sign-alternating. Between regimes 1 and 2 a hysteresis takes place. In region $G > 6700$ the oscillations become irregular (regime 3); a two-vortex flow structure is kept. At larger G values the motion map looks like a superposition of one-vortex and four-vortex structures (regime 4). A fragment of the typical picture of value S_1 change in time is shown in Fig. 4.16. In Fig. 4.17 a phase trajectory for this case is shown. The number of oscillations taking place in regions $S_1 > 0$ and $S_1 < 0$ changes in an irregular manner.

At $G > G_* = 10025$ the oscillations again become regular (regime 5; see Fig. 4.18). Thus, in the problem under consideration the region of irregular oscillations turns out to be restricted in terms of parameter G both from below and from above. Isoline cards, corresponding to points A, B, C in Fig. 4.18 are shown in Fig. 4.19 (a, b, c). The oscillations maintain a regular character in the whole studied region of Grashof number values (up to $G = 25000$). With the increase in G the oscillatory period decreases from 0.89 till 0.24 as G changes from 10030 to 20000.

4.4.   Convection with uniform heating from above

As mentioned in Section 2.3, free convection when the temperature gradient is directed vertically up is a phenomenon specific for a two-layer system. The instability of the steady state in this case is stipulated by considerable difference in both fluids' parameters, and as it follows from linear theory [49], [187] for convection emergence the lower fluid must have much larger heat diffusivity and coefficient of heat expansion than the upper fluid.

As in the case of heating from below, the steady state of fluid heated from above is stable at small temperature difference between horizontal cavity boundaries and loses stability at reaching any critical temperature difference. In a system of two infinite horizontal fluid layers the critical value of the Grashof number G depends on the wave number k of motion, function $G(k)$ having a minimum at any threshold G value, see Section 2.3.

The present section is devoted to the account of convective flows nonlinear results, arising in the system of horizontal fluid layers heated from above [55]. A system of equations (4.1)–(4.8) was solved; on the lateral boundaries of the calculation region conditions (4.18) were put; on solid horizontal boundaries we have

$$z = 1: \quad \psi_1 = \frac{\partial \psi_1}{\partial z} = 0, \quad T_1 = 1,$$

$$z = -a: \quad \psi_2 = \frac{\partial \psi_2}{\partial z} = 0, \quad T_2 = 0.$$

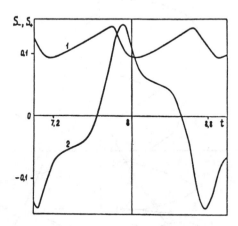

Figure 4.15. The change of flow structure in time at regular oscillations.

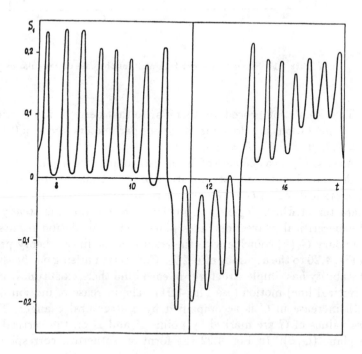

Figure 4.16. Irregular oscillations ($G = 9900$).

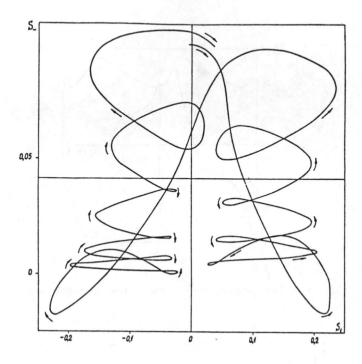

Figure 4.17. Phase trajectory fragment for the case of irregular oscillations.

The calculation procedure is similar to that described in Sections 4.1, 4.2. Calculations are done for the set of parameters $\kappa = \chi = 0.1$, $\nu = 1$, $\eta = 0.2$, $\beta = 0.01$, $P = 1$, $a = 1$.

As $G < G_*(k)$ a decrease of finite disturbances takes place irrespective of the initial state: the stabilization process leads to steady state. In supercritical region $G > G_*(k)$ as a result of a transitional process, the character of which depends on initial data, the limiting steady regime of supercritical convection takes place. The calculation results of the boundary $G_*(k)$ coincide with the results of the linear theory presented in Fig. 4.20 to the accuracy of 1–2 %. Convection arising on the threshold of stability has simple structure: in each fluid there exists one-vortex (on a vertical line) motion (see Fig. 4.21). The increase of motion intensity with increase in G is accompanied by a structural change. The first two values of G are marked by points A and B on the vertical section in plane (G, k). In Fig. 4.22 the form of isotherms, corresponding to

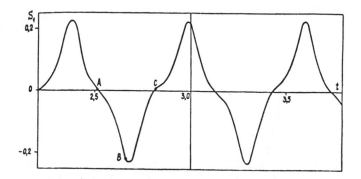

Figure 4.18. Regular oscillations ($G = 11000$).

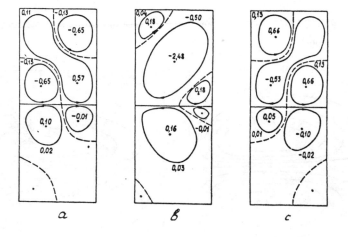

Figure 4.19. The change of stream lines picture at regular oscillations.

the motion 4.21 b is shown. As one can see from Fig. 4.21, with the increase in the Grashof number the convection intensity grows, a new vortex arises and gradually grows close to the upper plane of the first layer. At larger G (Fig. 4.21 d) this vortex motion becomes comparable in volume with the main motion.

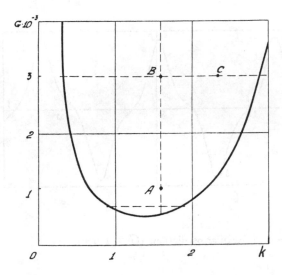

Figure 4.20. The neutral curve for the case of heating from above (linear theory).

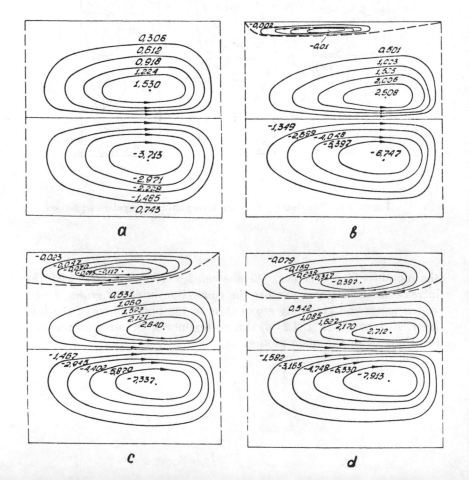

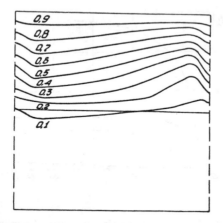

Figure 4.22. Isotherm picture (k=1.57; G=3000).

In Fig. 4.23 and 4.24 the change of convective motion intensity as we recede from the stability boundary is shown. The dependence of stream function extreme values on the Grashof number, presented in Fig. 4.23, is obtained for a fixed value of wave number $k = 1.57$ (vertical section in Fig. 4.20); curves 1 and 2 correspond respectively to the vortices of the upper and the lower fluids, close to the interface. Close to $G = G_*(k)$ function $\psi_m(G)$ is square which is typical for bifurcation with "soft" convection. Shaded curve 3 corresponds to an additional weak vortex arising in the upper layer.

Changing the region length $L = \pi/k$ at a fixed supercritical value of the Grashof number G it is possible to get a horizontal section of the instability region. Extreme values of $|\psi_m|$ depending on the wave number are shown in Fig. 4.24. The lower pair of curves concerns section $G = 700$, shown in Fig. 4.20; curves 1 and 2 correspond to upper and lower vortices. Both curves are located within the bounds between the points, corresponding to the stability boundaries according to the linear theory. The upper pair of curves refers to section $G = 3000$, also shown in Fig. 4.20 (line 3 corresponds to the upper fluid, line 4 - to the lower one). The intensity of motion turns out to be maximum at some value of k. At Grashof numbers close to the threshold, the variation of k does not lead to any visible change of flow structure (see Fig. 4.25). The situation changes, however, at larger values of supercriticity. For

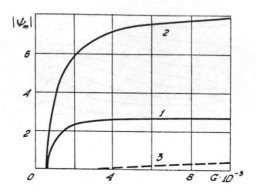

Figure 4.23. The change of maximum values of stream function module with the increase in the Grashof number.

example, at $G = 3000$ (the upper horizontal section in Fig. 4.20), the increase of k from $k = 1.57$, corresponding to Fig. 4.21 b, is accompanied by intensity increase of the secondary vortex. It can be seen if Fig. 4.21 b and Fig. 4.26, corresponding to $k = 2.35$ and $G = 3000$ (point C in Fig. 4.20) are compared.

The decrease of k from 1.57 leads to the dissapearance of the secondary vortex. The intensity maximum of motion as $G = 3000$ is displaced considerably to the left boundary of the instability region ($k = 0.9$) and is reached at the most simple motion structure. It is worth noting that at further decrease of k a longwave branch of curves is not realized. Apparently, in a longwave region one-vortex convective motion is unstable with respect to the break up of two vortices. A similar situation takes place in a supercritical convection in a flat horizontal layer heated from below [162] and in secondary convective flows in a vertical layer [56].

4.5.   Convection with longitudinal temperature gradient

In the previous sections of this chapter the variants of heating for which at moderately large values of the Grashof number the system possesses a mechanical reference state (the temperature gradient is directed perpendicularly to the interface) have been investigated. In this case convection

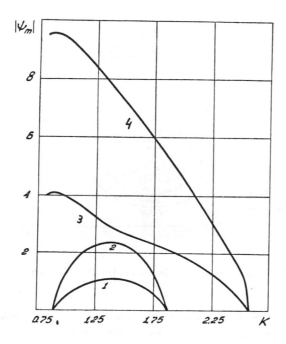

Figure 4.24. Dependence of maximum values of stream function module on the wave number.

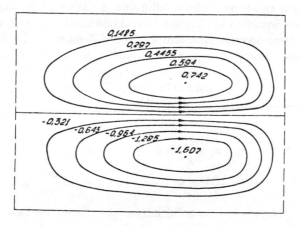

Figure 4.25. Stream lines picture for $k=1.125$; $G=700$.

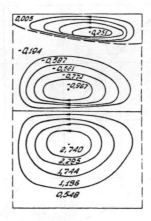

Figure 4.26. Stream lines picture for $k=2.35$; $G=3000$.

develops in a threshold manner as a result of the steady state instability. Another situation is realized as the temperature gradient is directed along the horizontal interface. At any small value of the Grashof number the existence conditions of mechanical steady state are violated and convective motion develops in the system.

1. Convection in a rectangular cavity. We shall consider convection in a two-layer system, filling up a cavity of rectangular cross section (see Section 4.1). Let the lateral boundaries be kept at constant nonequal temperatures (the temperature difference is θ). We shall use dimensionless variables in the same manner as in Section 1.2, choosing θ as the unit of temperature. Then equations (4.1)–(4.6) and boundary conditions (4.7), (4.8) retain their form. The conditions for the stream function on the solid boundaries also do not change:

$$x = 0, L: \quad \psi_m = \frac{\partial \psi_m}{\partial x} = 0 \quad (m = 1, 2),$$

$$z = 1: \quad \psi_1 = \frac{\partial \psi_1}{\partial z} = 0; \quad z = -a: \quad \psi_2 = \frac{\partial \psi_2}{\partial z} = 0.$$

(4.21)

For the temperature on solid lateral boundaries we have:

$$x = 0: \quad T_m = 1; \quad x = L: \quad T_m = 0 \quad (m = 1, 2).$$

(4.22)

We shall consider two types of conditions for the temperature on horizontal boundaries of the region: heat insulation:

$$z = 1: \quad \frac{\partial T_1}{\partial z} = 0; \quad z = -a: \quad \frac{\partial T_2}{\partial z} = 0 \qquad (4.23)$$

and ideal heat conductivity:

$$z = 1, -a: \quad T_m = 1 - \frac{x}{L}. \qquad (4.24)$$

Problem (4.1)–(4.8), (4.21)–(4.23) was solved for the first time in [176]. For the calculation of physical fields in the internal region an implicit scheme of variable directions method was used; the calculation of the stream function was done by the method of consecutive upper relaxation. To define the vortex on the interface a new iterative circle was introduced. The calculations were done on mesh 11×21 for a model system for which Grashof numbers coincided in both media and Prandtl numbers differed by a factor of two; $L = 2, a = 1$.

A similar problem for an air-water system is considered in [85]. Calculations are done for three values of layers' thicknesses ratio $a = 1; 3; 1/3$; in this case $L = (1+a)/2$ was chosen. The number of mesh nodes varied from 11×11 to 41×41 in each fluid. Let us give a more thorough account of the calculation results carried out in [167] for a system with ideally heat-conducting boundaries on the basis of calculating procedure described in Section 4.1. The uniform mesh 16×32 was used. The accuracy of the Poisson equation iteration for calculation of steady motions was 10^{-4}, and 10^{-7} for calculation of oscillatory motions. At oscillatory regimes investigation was based on the method, explained in Section 4.3.

Let us describe the results for water-silicone oil DC N 200 system (see Table 1; $a = 1; L = 2$). When heating from the lateral wall the mechanical steady state turns out to be impossible; a steady convective motion arises at any small Grashof numbers. An example of the flow map and the form of isotherms at small values of the Grashof number are presented in Fig. 4.27 a, b. The flow consists of one vortex in the upper fluid and a "two- storeyed" structure in the lower one, the motion of the fluid along the interface in both media being in one direction. The most intensive is the vortex in the upper layer (the intensity measure of the respective vortex is the extreme value of stream function). With increase in G the convection intensity and distortion of isotherms grow (Fig. 4.28 a, b). The dependence of extreme values of stream function $\psi_{m_{1,2}}$ on the Grashof number is presented in Fig. 4.29. Curve 1 corresponds to the vortex, situated in the upper layer and curves 2 and 3 correspond to the lower vortex (line 2 corresponds to the vortex, situated near the interface).

As $G \simeq 35000$ the steady motion becomes unstable and regular oscillations in the system develop. To characterize the oscillatory motion structure we shall use variables introduced by formulas (4.20). In Fig. 4.30 the dependence of $|S_1|$ on time is shown (the value S_1 is described by (4.20); see Section 4.3). In the process of oscillations in the upper fluid an additional vortex appears; it has the sign opposite to the sign of the "main" vortex. In the lower fluid a two-vortex structure with a periodically changing intensity is maintained. The hysteresis phenomena between the steady and oscillatory regimes have not been found. In the region of values $G \geq 5.5 \cdot 10^4$ the oscillations become non-periodical. The fragment of a typical evolution of $|S_1|$ in time is presented in Fig. 4.31.

In Fig. 4.32 a phase trajectory for the given case is presented. Variables S_1 and S_2 are negative and in the oscillation process do not change their sign.

As $L = 0.8$ the oscillatory circle bifurcates from a "three-storeyed" (one vortex is in the upper layer and two vortices are in the lower layer) steady motion only at $G \geq 310^5$. The maps of stream lines and isotherms for the steady motions as $L = 0.8$, corresponding to different values of the Grashof number are presented in Fig. 4.33.

2. Convection on a vertical flat plate. In [131] convection provoked by a heated plate has been investigated numerically and experimentally (see Fig. 4.34). A solid plate of finite length h (for the length unit, as before, the upper layer thickness is used) was located in the middle of the cavity (as $x = L/2$) and sunk in the lower fluid to the depth of h_1. On the plate constant flux Q_0 was kept. External cavity boundaries ($x = 0, L; z = 1, -a$) are assumed to be isothermal ($T = 0$) and solid.

The fluid motion and temperature distribution was assumed to be symmetric with respect to the transformation $x \rightarrow L-x$ which permitted us to make calculations for one half of region $0 \leq x \leq L/2$. The problem is described by a system of equations, presented in Section 4.1, but the conditions on the region boundaries, are now quite different.

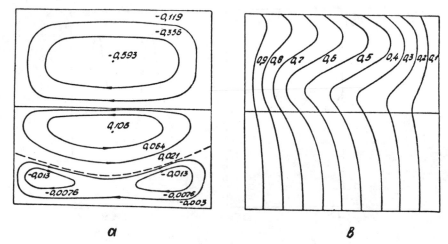

Figure 4.27. Isolines of stream function (a) and isotherms (b) for G=500.

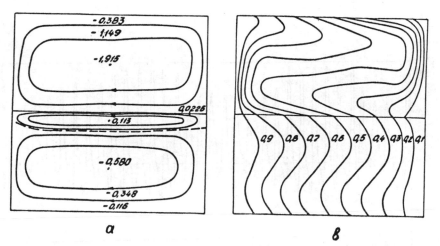

Figure 4.28. Isolines of stream function (a) and isotherms (b) for G=5000.

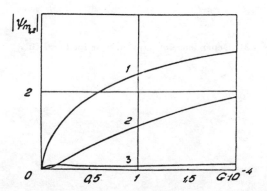

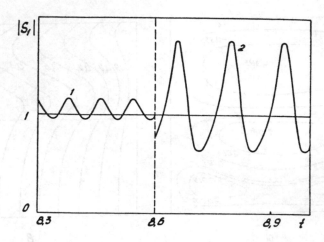

Figure 4.30. Dependence of $|S_1|$ on time for $G=4\cdot10^4$ (line 1) and $G=5\cdot10^4$ (line 2).

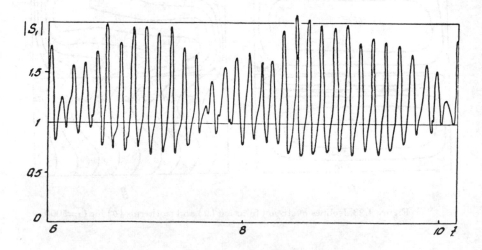

Figure 4.31. Dependence of $|S_1|$ on time for $G=6\cdot10^4$.

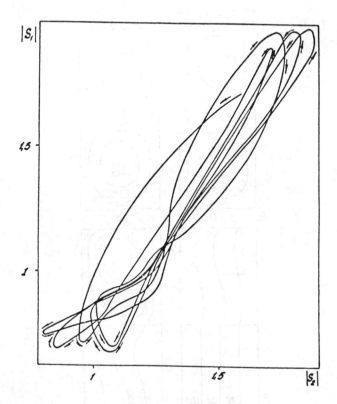

Figure 4.32. Phase trajectory for $G=6\cdot10^4$.

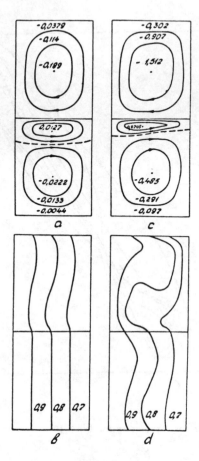

Figure 4.33. Stream lines and isotherms for G=500 (a, b) and G=10000 (c, d).

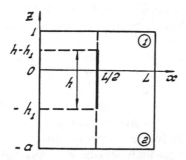

Figure 4.34. Geometry of the region.

$$z = 1: \quad v_1 = 0, \quad T_1 = 0,$$

$$z = -a: \quad v_2 = 0, \quad T_2 = 0,$$

$$x = 0, \quad z \geq 0: \quad v_1 = 0, \quad T_1 = 0,$$

$$x = 0, \quad z \leq 0: \quad v_2 = 0, \quad T_2 = 0,$$

$$x = \frac{L}{2}, \quad z \geq h - h_1: \quad v_{1,x} = 0, \quad \frac{\partial v_{1,y}}{\partial x} = 0, \quad \frac{\partial T_1}{\partial x} = 0,$$

$$x = L/2, \quad 0 \leq z \leq h - h_1: \quad v_1 = 0, \quad \frac{\partial T_1}{\partial x} = 1,$$

$$x = \frac{L}{2}, \quad -h_1 \leq z \leq 0: \quad v_2 = 0, \quad \frac{\partial T_2}{\partial x} = \kappa,$$

$$x = \frac{L}{2}, \quad -a \leq z \leq h_1: \quad v_{2,x} = 0, \quad \frac{\partial v_{2,y}}{\partial x} = 0, \quad \frac{\partial T_2}{\partial x} = 0.$$

The Grashof number is expressed through the problem parameters by

$$G = \frac{g \beta_1 Q_0 a_1^4}{\kappa_1 \nu_1^2}.$$

The problem has been solved by the method of meshes in initial variables v_m, p_m. The pressure was defined by means of the Poisson equations solution, obtained as a result of the operator *div* action on the motion equations, with the boundary conditions of Neumann type, which were found by projection of motion equations on the normal to the surface. The main calculations were done on meshes 50×25 (on the stabilization stage) and 70×33 (for the calculation of a steady flow); the mesh thickened near the solid boundaries and the interface and also in stagnation zones. An implicit scheme has been used; the derivaties on the space variables have been approximated by the central differences. To get the solution the implicit procedure, suggested in [175] has been used. The Poisson equation was solved on the multi-mesh method basis. The calculations were done for an air-water system; the h_1/h ratio had the values of 0.25; 0.5 and 0.75.

The calculation procedure permitted us to describe in detail the non-steady formation process of hydrodynamic and heat boundary layers on a heated plate, to study the evolution of temperature distribution on the plate surface, to calculate steady velocities and temperature fields in both media. The calculated profiles of temperature agree with experimental results.

3.   **Convection in a gap between horizontal cylinders.**   In [132] convection in a gap between concentric and eccentric horizontal cylinders with a constant and different temperature being kept is under consideration (see Fig. 4.35). The motion and temperature distribution are assumed to be symmetric with respect to the vertical plane, so the calculations were done for one half of the region. For the region of such geometry the application of the curvilinear coordinate system becomes natural. Constructing the mesh highly adapted to the shape of the gap the method proposed in [181] was used; the coordinate system was derived from the solution of certain elliptical problem in eigenvalues. Calculations were done for the value $R = 0.625$ and for three distance values between the centres of the cylinders: $\varepsilon = 0$ and $\varepsilon = \pm 0.625$. To calculate the convection between concentric cylinders mesh 33×25 has been used for both fluids. In the eccentric case cylinders of the regions filled up with different fluids are different; the region of a larger dimension was covered by mesh 33×33 and a smaller region — by mesh 33×21. The calculations are done for a series of systems including air-water and silicone oil-water systems. Calculations permitted to obtain the stream lines and isotherms in a wide range of Grashof numbers, to investigate the distribution of the local Nusselt number along the surface of an internal and external cylinder, to get the approximate empiric formulas for the dependences between the middle Nusselt number and the Grashof number. The comparison of flow patterns and heat fields for different fluids pairs has been carried out.

4. **Convection between vertical cylinders of finite height.** In the previous sections we discussed two-dimensional flows. Let us consider now the convection of axially symmetric type in vertical cylinders of finite height filled with two fluids. Inside a cylinder a concentric cylin-

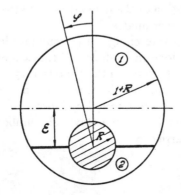

Figure 4.35. A gap between horizontal eccentric cylinders.

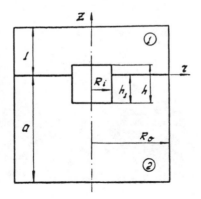

Figure 4.36. The system of vertical cylinders.

der of smaller dimensions is placed (see Fig. 4.36) [97]. The cylinders
are kept at constant nonequal temperatures.

The core of the approach is the fact that the internal cylinder is
supposed to be filled up with a fluid of infinite viscosity and heat conductivity. This condition ensures vanishing of velocity, and temperature
constancy on its surface. This approach allows us not to state boundary conditions on the internal cylinder surface; the boundary conditions

are set only on the interface $z = 0$ and on the surfaces of the external cylinder $(z = 1, z = -a, r = R_0)$. The convection equations are written with the consideration of the dependence of viscosities η_m and heat conductivities κ_m on coordinates.

Non-uniform meshes thickening in the boundary layers are used. The calculations are being done on mesh 28×23 (in each region) for the series of two-layer systems. The sink depth of internal cylinder h_1/h is varied. The main attention is paid to the flow pattern, velocity profiles and heat flux distribution on the internal cylinder. For the dependence of the average Nusselt number on the Rayleigh number, comparison with experiment [173] is carried out.

CHAPTER 5

Well-Developed Thermocapillary Convection

A two-layer approach turns out to be the most prospective for thermocapillary convection problems, in contrast with thermogravitational convection stipulated by surface effects. A consecutive calculation of thermal and hydrodynamic interaction between the media permits one to eliminate empirical parameters usage (such as the Biot number).

In this chapter on the basis of full nonlinear equations the finite-amplitude convection regimes caused by thermocapillary mechanism action are investigated. The computing capacity compels us to investigate convective motions of a respectively simple space structure only (two-dimensional or axially symmetric), which are realized in the cavities with a moderate sides ratio and not too large motion velocities. In real systems of horizontal layers motions of much more complicated space structure can be realized including cells on different scales [96]. Convection may cause the so-called "interphase turbulence", leading to an essential increase of transfer coefficients [150]. We are not going to consider these phenomena, for their investigation in terms of nonlinear theory is not yet complete.

In Section 5.1 the boundary value problem is solved under assumption of a flat nondeformable interface between the media; in Sections 5.2 and 5.3 the influence of two different factors: distortion of the interface and Archimedes buoyancy is investigated. The contents of these paragraphs is based on [103 - 106, 109, 113]. The review of the papers devoted to the investigation of thermocapillary convection with a temperature gradient directed along the interface is given in Section 5.4. In Section 5.5 a drop motion under the action of temperature gradient stipulated by a thermocapillary effect is studied.

5.1. Convection in a system with a flat interface

1. Statement of the problem and method of solution. As in the previous chapter, we shall consider convection in a horizontal cylinder of a rectangular cross section ($0 \leq x \leq l$, $-a_2 \leq z \leq a_1$). The interface is assumed to be flat and nondeformable: $z = 0$. Solid horizontal boundaries are kept at constant and nonequel temperatures; heating both from below and above is possible. On solid vertical walls temperature distribution, ensuring mechanical reference state, is maintained. The coefficient of surface tension linearly depends on temperature. Archimedes force in this section is not taken into consideration.

Passing on to dimensionless variables in the same way as in Section 1.2 and introducing the stream function and vortex, we get the following system of equations:

$$\frac{\partial \varphi_m}{\partial t} + \frac{\partial \psi_m}{\partial z}\frac{\partial \varphi_m}{\partial x} - \frac{\partial \psi_m}{\partial x}\frac{\partial \varphi_m}{\partial z} = c_m \Delta \varphi_m,$$

$$\Delta \psi_m = -\varphi_m, \tag{5.1}$$

$$\frac{\partial T_m}{\partial t} + \frac{\partial \psi_m}{\partial y}\frac{\partial T_m}{\partial x} - \frac{\partial \psi_m}{\partial x}\frac{\partial T_m}{\partial y} = \frac{d_m}{P}\Delta T_m, \quad m = 1, 2.$$

Here $c_1 = d_1 = 1$, $c_2 = 1/\nu$, $d_2 = 1/\chi$. On the interface the following conditions are fulfilled:

$$z = 0 : \psi_1 = \psi_2 = 0, \quad \frac{\partial \psi_1}{\partial z} = \frac{\partial \psi_2}{\partial z},$$

$$T_1 = T_2, \quad \kappa\frac{\partial T_1}{\partial z} = \frac{\partial T_2}{\partial z}, \tag{5.2}$$

$$\varphi_2 = \eta\varphi_1 + Mr\frac{\partial T_1}{\partial x}. \tag{5.3}$$

Two variants of defining the temperature on solid boundaries are considered. For heating from below:

$$z = 1 : \psi_1 = \frac{\partial \psi_1}{\partial z} = 0, \quad T_1 = 0;$$

$$z = -a : \psi_2 = \frac{\partial \psi_2}{\partial z} = 0, \quad T_2 = 1; \tag{5.4}$$

$$x = 0, L : \psi_1 = \frac{\partial \psi_1}{\partial x} = 0, \quad T_1 = \frac{1 - z}{1 + \kappa a} \quad (z > 0),$$

$$\psi_2 = \frac{\partial \psi_2}{\partial x} = 0, \quad T_2 = \frac{1 - \kappa z}{1 + \kappa a} \quad (z < 0). \tag{5.5}$$

and for heating from above:

$$z = 1 : \psi_1 = \frac{\partial \psi_1}{\partial z} = 0, \quad T_1 = 1;$$

$$z = -a : \psi_2 = \frac{\partial \psi_2}{\partial z} = 0, \quad T_2 = 0; \tag{5.6}$$

$$x = 0, L : \psi_1 = \frac{\partial \psi_1}{\partial x} = 0; \quad T_1 = \frac{z + \kappa a}{1 + \kappa a} \quad (z > 0),$$

$$\psi_2 = \frac{\partial \psi_2}{\partial x} = 0; \quad T_2 = \frac{\kappa(z + a)}{1 + \kappa a} \quad (z < 0).$$

(5.7)

Along with the convection in an enclosed rectangular cavity we shall consider a two-dimensional space-periodic convection in a system of horizontal layers. In this case boundary conditions (5.4) and (5.6) are replaced by the periodic conditions:

$$\psi_1(2L, z) = \psi_1(0, z), \quad \varphi_1(2L, z) = \varphi_1(0, z),$$
$$T_1(2L, z) = T_1(0, z) \quad (z \geq 0),$$
$$\psi_2(2L, z) = \psi_2(0, z), \quad \varphi_2(2L, z) = \varphi_2(0, z),$$
$$T_2(2L, z) = T_2(0, z) \quad (z \leq 0);$$

(5.8)

here $2L$ is the space period of a convective flow. Steady motions being symmetric, their calculation may be carried out on one half of a space period with the statement of conditions on convective cells boundaries:

$$x = 0, L : \psi_1 = \varphi_1 = 0, \quad \frac{\partial T_1}{\partial x} = 0 \quad (z \geq 0),$$

$$x = 0, L : \psi_2 = \varphi_2 = 0, \quad \frac{\partial T_2}{\partial x} = 0 \quad (z \leq 0).$$

(5.9)

The solution of the boundary-value problem was carried out with the help of the finite-difference method. The computing procedure and the form of initial disturbances are similar to those discribed in Sections 4.1 and 4.2. The calculation of temperature on the interface was carried out using formula of the second order accuracy (4.19). In contrast with the thermogravitational convection, on the interface the vortex in the upper fluid was calculated by formula

$$\varphi_1(x, 0) = \frac{-2[\psi_2(x, -\Delta z) + \psi_1(x, \Delta z)]}{(\Delta z)^2 (1 + \eta)} - Mr \frac{1}{1 + \eta} \frac{\partial T_1}{\partial x}(x, 0),$$

which can be deduced in analogy with the Thom formula, and in the lower fluid by formula (5.3).

The investigation of nonsteady convection regimes being performed, the same verifications were used as in Section 4.3. Moreover, the formulas of the first order accuracy for the temperature calculation on the interface was used and the way of mesh nodes pass at the Poisson equation iteration was varied. The structure of nonsteady motions in all cases was kept and threshold number Mr and oscillations amplitude change was not considerable. For example, the first order accuracy formulas for the temperature on the interface being used, the critical Marangoni number was reduced by 10%. The time step decreasing twice, in the region of well-developed convection oscillations amplitude decrease is not more than 3%.

Because of a large number of problem parameters we shall consider only some typical cases leading to qualitatively new results.

2. Development of monotonous instability. One of the important predictions of the linear stability theory of a horizontal layers system (Section 2.5) is the conclusion that at $\chi < 1$ and $a < 1/\sqrt{\chi}$ when heating from the side of the upper fluid monotonous instability takes place. The opposite way of heating being used, absolute steady state stability takes place. Numerical calculations show that this conclusion is true for an enclosed cavity too. As an example we shall consider the system with coefficients of heat conductivity and heat diffusivity for the upper fluid twice as smaller as for the lower one ($\kappa = \chi = 0.5$; $\eta = \nu = 1$; $P = 1$; $L = 2$; $a = 1$).

Thermogravitational convection in this system was investigated in Chapter 4. Let us recall of the main results. As has been noted in Section 4.2, the Rayleigh numbers separately defined for the upper and the lower fluids, differ by a factor of four because of temperature gradients and heat diffusivities difference. Therefore as $G \neq 0$ with increase of temperature difference critical conditions for thermogravitational convection generation in the upper fluid are achieved earlier than in the lower one. As a result at exceeding of the critical Grashof number convection stipulated by Archimedes force develops only in the upper fluid, and in the lower one there exists a weak flow induced by stresses on the interface.

For thermocapillary convection ($Mr \neq 0$, $G = 0$) the situation is qualitatively different. The temperature disturbance on the interface causes the motion simultaneously in both fluids. The dependence of stream function maximum value ψ_{1m} on $Mr > 2500$ number at $a = 1$ and $L = 2$ (heating is given by formulas (5.6) and (5.7)) and typical isolines pattern is presented in Fig. 5.1. As a result of the reference state stability loss as $Mr > 2500$, a steady convective flow of a four-vortex structure symmetric with respect to vertical axis develops:

$$\psi_m(x, z) = -\psi_m(L - x, z),$$
$$T_m(x, z) = T_m(L - x, z), \quad m = 1, 2. \tag{5.10}$$

Besides, in this case ($\eta = \nu = a = 1$) a specific symmetry: $\psi_1(x, z) = -\psi_2(x, -z)$, $0 < z < 1$ exists. The vortices have a typical

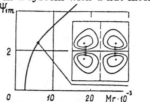

Fig. 5.1. The amplitude curve and isolines picture for a model system.

for thermocapillary convection "flattened" form, the intensive motion is realized near the interface. Let us notice that both two- and four- vortex initial disturbance leads to one and the same stationary regime.

With heating from below given by (5.4) and (5.5) the system keeps a mechanical reference state in the whole range of values $Mr < 2 \cdot 10^5$. This situation can be explained in the following way. Let the positive temperature disturbance appear near the interface. The thermocapillary effect causes the flow of a cold fluid from above and the flow of a hot fluid from below. In both fluids the flow has an identical intensity. However, owing to heat diffusivity media difference, the up rising hot fluid becomes quickly cool, and the down going cold fluid becomes slowly warm leading to the decrease of disturbance.

Consider now a system of real fluids, water-silicone oil DC N 200 (see Table 1); $L = 2$; $a = 1$ [63]. The linear stability theory of horizontal layers for this system is constructed in Section 2.5. In case of heating from the side of the first fluid on exceeding the threshold Mr number the steady state becomes unstable and a thermocapillary convection arises in the system. In Fig. 5.2 the dependence of stream function maximum modulus $|\psi_m|$ on Mr number (curve a corresponds to the boundary conditions (5.9) and curve b — to conditions (5.6) and (5.7)).

Note that as far as the coefficients of fluids dynamic and kinematic viscosities are close, the maximum motion intensity curves in the first and the second media practically coincide on the graph scale.

Let us consider the case of convection in the system of infinite horizontal layers (boundary conditions (5.4) and (5.9)). The patterns of stream function isolines and isotherms are presented in Fig. 5.3 a, b at $Mr = 2500$. One can see that the flow velocity is at maximum close to the interface. Because of different heat conductivities the isotherms turn out to be concentrated mainly in the first fluid. With increase of Mr, the steady motion loses stability, and regular oscillations develop in the system. In Fig. 5.4 the dependence of $|S_1(t)|$ (see formulas 4.20) for $Mr = 5000$ is presented. The change of flow structure in the oscillations process is not considerable. The typical stream function isolines corresponding to points A and B of Fig. 5.4 are presented in Fig. 5.5 a, b.

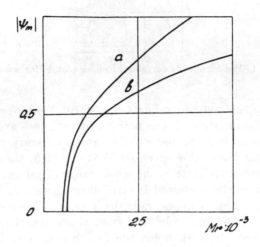

Figure 5.2. The amplitude curves for a water-silicone oil N 200 system.

Further increase of Mr number is accompanied by the complication of oscillations (see Fig. 5.6) which in the investigated range of parameter Mr (up to $Mr = 9000$) keep a regular character.

For the cavity with all-round solid boundaries (conditions (5.4), (5.5)) the convection threshold increases (see Fig. 5.2). The steady motion structure is similar to the pattern presented in Fig. 5.3 a. When heating from the side of the second fluid thermocapillary convection does not arise.

3. Development of oscillatory instability. As shown in §2.5, the onset of thermocallary convection in a two-layer system may have an oscillatory character. As an example we shall present calculated results for a system with parameters $\eta = \nu = 0.5$; $\kappa = \chi = 1$; $P = 1$; $a = 1$ in which thermocapillary oscillations are the only source of instability. Let $L = 2.5$. When heating from below, given by (5.4) and (5.5), mechanical steady state is kept at $Mr < Mr_1 \simeq 15700$. In subcritical region ($Mr = 15000$) the initial four-vortex disturbance vanishes in oscillatory manner. As $Mr > Mr_1$ the convective oscillations further denoted by $S_1(t)$ and $S_2(t)$ (see 4.20) and the value $\psi_+ = \max_t \max_{x,z} |\psi_1(x, z, t)|$ arise. The motion is a four-vortex, symmetric with respect to the vertical axis $x = L/2$ structure, for which $S_1 = -S_2$. The dependence of $S_1(t)$ for different values of Mr is shown in Fig. 5.7. Close to the threshold the oscillations

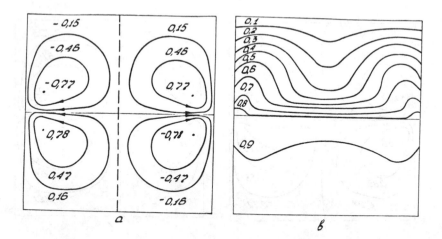

Figure 5.3. Isoline pictures of stream function (a) and isotherms (b).

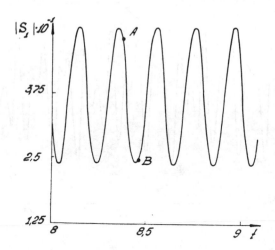

Figure 5.4. Harmonic oscillations.

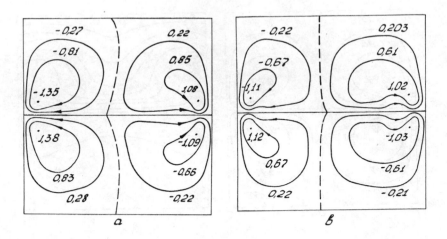

Figure 5.5. Isoline pictures of stream function for oscillatory convection.

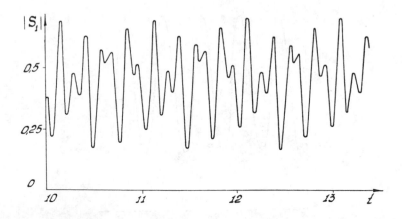

Figure 5.6. Periodical oscillations of complicated form.

are close to sinusoidal; their period is $\tau \simeq 0.46$, the amplitude changes with the increase in $Mr - Mr_1$ according to square-root law. With the increase in Mr the oscillations become essentially nonlinear, their period grows. In Fig 5.8 the dependence of τ^{-2} on Mr (line 1) is presented. In the range $22500 < Mr < Mr_2 = 25000$ the oscillation period is well approximated by formula

$$\tau^{-2} = 0.62(Mr_2 - Mr). \qquad (5.11)$$

As $Mr > Mr_2$ a steady convective flow arises. Dependence (5.11) allows to suppose that as $Mr = Mr_2$ the oscillatory circle transforms into a separatrix of non-rough saddle point (saddle-node) [6]. There is no hysteresis between steady and oscillatory regimes. Note that a similar situation was observed in the problem of conducting fluid convection in a magnetic field [171].

In Fig. 5.8 the dependence of ψ_+ on Mr for an oscillatory regime (line 2) and ψ_{1m} for steady regime (line 3) are shown. Variables $q_- = \lim_{Mr \to Mr_2 - 0} \psi_+$ and $q_+ = \lim_{Mr \to Mr_2 + 0} \psi_{1m}$ differ appreciably. Variable q_+ is close to the average value of $\psi_{1m}(t)$ on the nonlinear oscillations period.

When heating from above this system retains stability at least up to $Mr = 2 \cdot 10^5$.

A more complicated sequence of transitions is encountered when the ratio of cavity sides is $L = 2$. As in the case of $L = 2.5$, as a result of a steady state instability the oscillatory motion symmetric with respect to the vertical axis arises ($S_1 = -S_2$). With the increase in Mr the bifurcation of period doubling, accompanied by a break up of flow symmetry takes place. In the course of oscillation process both a change of motion intesity and a mutual displacement of vortices with an opposite rotation direction takes place. Dependence $S_1(t)$ gets a more complicated character (see Fig. 5.9). Equality $S_1 = -S_2$ is violated; the typical phase trajectory in plane (S_1, S_2) is shown in Fig. 5.10. A further insrease of Mr leads to a steady convection.

5.2. Convection in a system with a curved interface

1. Statement of the problem and calculation method. In this section the calculation of thermocapillary convective motions, is carried out, a curved interface stipulated by the contact angle being taken into account [106, 109]. As noted in Section 1.2 this effect is essential under microgravity conditions. The boundary form $z = h(x)$ is determined from the normal stresses balance condition, which in a dimensionless

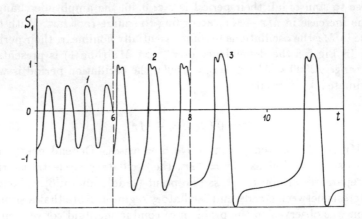

Figure 5.7. Thermocapillary oscillations form for $Mr=1.8\cdot10^4$ (line 1); $2.25\cdot10^4$ (2); $2.4\cdot10^4$ (3); $L=2.5$.

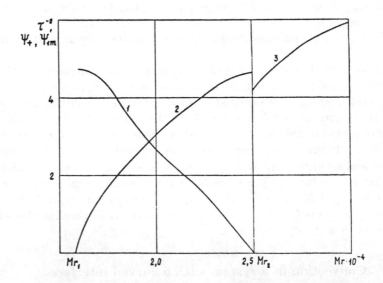

Figure 5.8. Dependence of oscillatory and stationary regimes characteristics on Mr number.

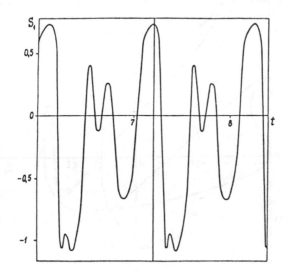

Figure 5.9. Thermocapillary oscillations form as $Mr=2.5{\cdot}10^4$, $L=2$.

form has the form (1.24). On the region boundaries equalities $h'(x) = \pm \mathrm{ctg}\gamma$ are fulfilled, where γ is a contact angle.

In this approach the interface shape $h(x)$ is fixed and does not depend on time. Particularly, under microgravity conditions ($r_c \rightarrow \infty$) the interface has the shape of the circle arch [79]:

$$h(x) = \frac{L}{2}\mathrm{tg}\gamma - \sqrt{\left(\frac{L}{2}\mathrm{tg}\gamma\right)^2 + Lx - x^2} \qquad (5.12)$$

(L is the horizontal cavity size). On the interface the normal velocity components vanish and the continuity conditions of tangential velocity components, velocity, temperature, heat fluxes and tangential stresses are fulfilled:

$$z = h(x): \quad \psi_1 = \psi_2 = 0, \quad \frac{\partial \psi_1}{\partial n} = \frac{\partial \psi_2}{\partial n}, \quad T_1 = T_2, \qquad (5.13)$$

$$\kappa\frac{\partial T_1}{\partial n} = \frac{\partial T_2}{\partial n}, \quad \sigma_{2,n\tau} = \sigma_{1,n\tau} - Mr\frac{\partial T_1}{\partial \tau},$$

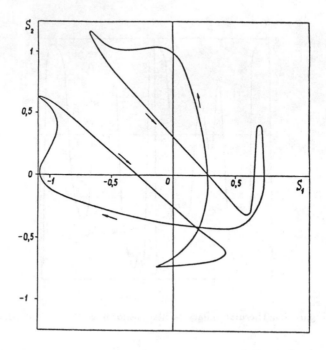

Figure 5.10. Phase trajectory for thermocapillary oscillations ($Mr=2.5\cdot10^4$, $L=2$).

where

$$\frac{\partial}{\partial n} = n_i \frac{\partial}{\partial x_i}, \sigma_{m,n\tau} = \sigma_{m,ik} n_i \tau_k,$$

n_i, τ_k are normal and tangential vectors to the interface, $\sigma_{m,ik}$ is the viscous stress tensor of m-th fluid.

As in 3 Section 1.2 for the calculation of convection in the case of a curved interface it is convenient to pass on curvilinear coordinate system. Let us introduce the coordinate system:

$$X = x, \mathcal{Z} = \begin{cases} 1 + \frac{z-1}{1-h(x)}, & z \geq h(x), \\ a\left[\frac{z+a}{a+h(x)} - 1\right], & z \leq h(x), \end{cases} \qquad (5.14)$$

in which equation of the interface is $Z = 0$. Equations for stream function ψ_m, vortex φ_m and temperature T_m take the form:

$$\frac{\partial \varphi_m}{\partial t} + F_m \left(\frac{\partial \psi_m}{\partial Z} \frac{\partial \varphi_m}{\partial X} - \frac{\partial \psi_m}{\partial X} \frac{\partial \varphi_m}{\partial Z} \right) = c_m D_m \varphi_m$$

$$D_m \psi_m = \varphi_m,$$

$$\frac{\partial T_m}{\partial t} + F_m \left(\frac{\partial \psi_m}{\partial Z} \frac{\partial T_m}{\partial X} - \frac{\partial \psi_m}{\partial X} \frac{\partial T_m}{\partial Z} \right) = \frac{d_m}{P} D T_m,$$

$$c_1 = d_1 = 1, \quad c_2 = \frac{1}{\nu}, \quad d_2 = \frac{1}{\chi},$$

$$F_1 = \frac{1}{1-h}, \quad F_2 = \frac{1}{1+ha^{-1}}, \tag{5.15}$$

$$D_m = \frac{\partial^2}{\partial X^2} + 2A_m \frac{\partial^2}{\partial X \partial Z} + B_m \frac{\partial^2}{\partial Z^2} + C_m \frac{\partial}{\partial Z}, \quad m = 1, 2$$

$$A_1 = (Z-1)h'F_1, \quad B_1 = A_1^2 + F_1^2,$$

$$C_1 = (Z-1)(F_1 h'' + 2h^{12}F_1^2), \quad A_2 = -(Z+a)a^{-1}h'F_2,$$

$$B_2 = A_2^2 + F_2^2, C_2 = (Z+a)a^{-1}(-F_2 h'' + 2a^{-1}h^{12}F_2^2).$$

Here differentiation by X is designated by prime.
The conditions on the interface are:

$$Z = 0: \quad \psi_1 = \psi_2 = 0, F_1 \frac{\partial \psi_1}{\partial Z} = F_2 \frac{\partial \psi_2}{\partial Z},$$

$$T_1 = T_2, \kappa F_1 \frac{\partial T_1}{\partial Z} = F_2 \frac{\partial T_2}{\partial Z}, \tag{5.16}$$

$$\varphi_2 - \eta \varphi_1 = Mr E^{1/2} \frac{\partial T_1}{\partial X} - 2h'' E F_1 \frac{\partial \psi_1}{\partial Z} (\eta - 1),$$

where $E = (1 + h^{12})^{-1}$.
The boundary conditions on solid walls:

$$X = 0, L: \quad \psi_m = \frac{\partial \psi_m}{\partial X} = 0,$$

$$Z = 1: \psi_1 = \frac{\partial \psi_1}{\partial Z} = 0, \tag{5.17}$$

$$Z = -a: \psi_2 = \frac{\partial \psi_2}{\partial Z} = 0 \quad (m = 1, 2).$$

Two variants of defining the temperature on solid boundaries, corresponding to heating from the side of the second fluid are considered:

$$Z = 1: \quad T_1 = 0, \quad Z = -a : T_2 = 1;$$

$$X = 0, \; L : T_1 = \frac{1 - Z}{1 + \kappa a} \; (Z \geq 0),$$

$$T_2 = \frac{1 - \kappa Z}{1 + \kappa a} \; (Z < 0) \tag{5.18}$$

or from the side of the first fluid:

$$Z = 1: \quad T_1 = 1; \quad Z = -a : T_2 = 0;$$

$$X = 0, \; L : T_1 = \frac{Z + \kappa a}{1 + \kappa a} \; (Z > 0),$$

$$T_2 = \frac{\kappa(Z + a)}{1 + \kappa a} \; (Z < 0). \tag{5.19}$$

As in the previous section, the problem was solved by the method of mesh approximation; equations and boundary conditions were approximated using a uniform mesh:

$$X_i = i\Delta X, i = 0, 1, \ldots, I, \Delta X = \frac{L}{I};$$

$$Z_j = j\Delta Z_1, j = 0, 1, \ldots, J_1, \Delta Z_1 = \frac{1}{J_1} \; (Z > 0);$$

$$Z_j = -a + j\Delta Z_2, j = 0, 1, \ldots, J_2, \Delta Z_2 = \frac{a}{J_2}(Z < 0).$$

Let us write the finite-difference analogs of the equations:

$$\varphi_{i,j}^{n+1} = \varphi_{i,j}^n + \Delta t \bigg\{ \frac{c_m}{(\Delta X^2)} \Delta \varphi_{i,j}^n + \frac{A_m c_m}{2\Delta X \Delta Z_m}(\varphi_{i+1,j+1}^n -$$

$$\varphi_{i-1,j+1}^n - \varphi_{i+1,j-1}^n + \varphi_{i-1,j-1}^n) + \frac{B_m c_m}{(\Delta Z_m)^2} \Delta \varphi_{i,j}^n +$$

$$\frac{C_m c_m}{2\Delta Z_m}(\varphi_{i,j+1}^n - \varphi_{i,j-1}^n) - \frac{F_m}{4\Delta X \Delta Z_m}[(\psi_{i,j+1}^n - \psi_{i,j-1}^n)(\varphi_{i+1,j}^n -$$

$$\varphi_{i-1,j}^n) - (\psi_{i+1,j}^n - \psi_{i-1,j}^n)(\varphi_{i,j+1}^n - \varphi_{i,j-1}^n)] \bigg\}; \tag{5.20}$$

$$\psi_{i,j}^{n,s+1} = \psi_{i,j}^{n,s} + \omega\{[\varphi_{i,j}^{n,s}(\Delta X)^2(\Delta Z_m)^2 + (\psi_{i+1,j}^{n,s} +$$

$$\psi_{i-1,j}^{n,s+1})(\Delta Z_m)^2 + \frac{A_m}{2}(\psi_{i+1,j+1}^{n,s} - \psi_{i-1,j+1}^{n,s+1} -$$

$$\psi_{i+1,j-1}^{n,s+1} + \psi_{i-1,j-1}^{n,s+1})\Delta X\Delta Z_m + B_m(\psi_{i,j+1}^{n,s} +$$

$$\psi_{i,j-1}^{n,s+1})(\Delta X)^2 + \frac{C_m}{4[(\Delta Z_m)^2 + (\Delta X)^2 B_m]} \times$$

$$\times (\psi_{i,j+1}^{n,s} - \psi_{i,j-1}^{n,s+1})(\Delta X)^2(\Delta Z_m)] - \psi_{i,j}^{n,s}\}; \qquad (5.21)$$

$$T_{i,j}^{n+1} = T_{i,j}^n + \Delta t\Big\{\frac{d_m}{P(\Delta X)^2}\Delta T_{i,j}^n +$$

$$\frac{A_m}{2\Delta X\Delta Z_m}(T_{i+1,j+1}^n - T_{i-1,j+1}^n - T_{i+1,j-1}^n + T_{i-1,j-1}^n) + \frac{B_m}{(\Delta Z_m)^2}\Delta T_{i,j}^n +$$

$$\frac{C_m}{2\Delta Z_m}(T_{i,j+1}^n - T_{i,j+1}^n) - \frac{F_m}{4\Delta X\Delta Z_m}[(\psi_{i,j+1}^{n+1} - \psi_{i,j-1}^{n+1}) \times$$

$$(T_{i+1,j}^n - T_{i-1,j}^n) - (\psi_{i+1,j}^{n+1} - \psi_{i-1,j}^{n+1})(T_{i,j+1}^n - T_{i,j-1}^n)]\Big\}, \quad m = 1, 2.$$

$$(5.22)$$

Here n is the time layer number and s is the iteration number. In notation for $\varphi_{m,ij}$, $\psi_{m,ij}$, $T_{m,ij}$ index m for brevity is ommitted.

The temperature on the interface is defined by

$$(T_1)_{i,0} = (T_2)_{i,J_2} = \frac{\kappa(F_1)_i(T_1)_{i,1}\Delta Z_2 + (F_2)_i(T_2)_{i,J_2-1}\Delta Z_1}{\kappa(F_1)_i\Delta Z_2 + (F_2)_i\Delta Z_1},$$

derived from boundary conditions (5.16). The vortex calculation on the interface is carried out in the following manner. For stream function values in the nearest to the interface nodes expansions in Taylor series up to the second order including the first condition from (5.16) are written

$$(\psi_1)_{i,1} = \frac{(\partial\psi_1)_{i,1}}{\partial Z}\Delta Z_1 + \frac{1}{2}\frac{(\partial^2\psi_1)_{i,1}}{\partial Z^2}(\Delta Z_1)^2,$$

$$(\psi_2)_{i,J_2-1} = -\frac{(\partial\psi_2)_{i,J_2-1}}{\partial Z} + \frac{1}{2}\frac{(\partial\psi_2)_{i,J_2-1}}{\partial Z^2}(\Delta Z_2)^2.$$

The variables $\partial^2\psi_m/\partial Z^2$ on the interface are expressed through the vortex values:

$$\frac{\partial^2 \psi_m}{\partial Z^2} = B_m^{-1}\left(-\varphi_m - 2A_m \frac{\partial^2 \psi_m}{\partial X \partial Z} - C_m \frac{\partial \psi_m}{\partial Z}\right).$$

Using the second condition from (5.16) and eliminating $\partial\psi_m/\partial Z$ from the obtained equations, we get a correlation which connects φ_1 and φ_2 on the interface:

$$\frac{1}{2}(\Delta Z_1)^2 (q_1)_i E_i F_{1,i}^{-2}(\varphi_1)_{i,0} + \frac{1}{2}(\Delta Z_2^2)(q_2)_i E_i F_{2,i}^{-2}(\varphi_2)_{i,J_2-1} =$$

$$- [(q_1)_i(\psi_1)_{i,1} + (q_2)_i(\psi_2)_{i,J_2-1}] +$$

$$(q_1)_i(\Delta Z_1)^2 F_{1,i}^{-1}(h')_i E_i \left(\frac{\partial^2 \psi_1}{\partial X \partial Z}\right)_{i,0} +$$

$$(q_2)_i(\Delta Z_2)^2 F_{2,i}^{-1}(h')_i E_i \left(\frac{\partial^2 \psi_2}{\partial X \partial Z}\right)_{i,J_2},$$

where

$$(q_m)_i = F_{m,i}\left\{\Delta Z_m\left[1 + \frac{1}{2}(-1)^m C_{m,i} E_i F_{m,i}^{-2}\Delta Z_m\right]\right\}^{-1},$$

$$h_i = h(X_i), \quad E_i = E(X_i), \quad (C_m)_i = C_m(X_{i,0}), \quad (F_m)_i = F_m(X_{i,0});$$

mixed derivatives from ψ_m are approximated by the differences, one-hand side by Z. For the second correlation allowing to define the vortex values on the interface, the last formula from (5.16) is used.

On solid boundaries the vortex is defined by

$$(\varphi_1)_{i,J_1} = -\frac{2}{(\Delta Z_1)^2}(\psi_1)_{i,J_1-1}(F_1)_i^2,$$

$$(\varphi_2)_{i,0} = -\frac{2}{(\Delta Z_2)^2}(\psi_2)_{i,1}(F_2)_i^2,$$

$$(\varphi_m)_{0,j} = -\frac{2}{(\Delta X)^2}(\psi_m)_{1,j},$$

$$(\varphi_m)_{I_m,j} = -\frac{2}{(\Delta X)^2}(\psi_m)_{I_m-1,j}; \quad m = 1, 2.$$

For the case of a flat interface this algorithm coincides with that of Section 5.1. Calculations have been done on mesh 16×32.

2. Calculation results. In the case of a flat interface the conditions of heating (5.18), (5.19) ensure mechanical reference state in the system, which at any value of Mr number $Mr = Mr_*$ loses stability in a monotonous or oscillatory manner (see §5.1). A distortion of the interface leads to the fact that mechanical steady state turns out to be impossible. Temperature heterogeneity along the interface leads to the generation of a steady motion at any small Mr values (see 2 Section 3.1). Let us discuss for definiteness the case of $\gamma < 90°$. When heating from the side of the second fluid the temperature on the interface is maximum in the cavity centre. It leads to the development of the motion with the form schematically shown in Fig. 5.11 (structure a). When the opposite variant of heating is used, the motion of type b is realized.

Let us suppose first that the motion structure stipulated by heterogeneous temperature distribution on the interface, coincides with the motion form which is realized in the case of a flat interface as a result of monotonous instability. Then the instability of the motion close to $Mr = Mr_*$ does not arise and its amplitude continuously grows. The situation is different if in the case of a flat interface instability of the steady state leads to the motion of a type different from the one arising because of the interface distortion. In this situation the transitions between the flows of different structure are possible, as Mr number changes.

As an example we shall consider a system with the following parameters $\kappa = \chi = 0.5$; $\eta = \nu = P = a = 1$; $L = 2$. As noted in Section 5.1, in the case of a flat interface ($\gamma = 90°$) when heating from above mechanical reference state loses stability at $Mr \simeq 2500$ with respect to the monotonously growing disturbances. It leads to the onset of a steady convective motion of structure a. In Fig. 5.11 for this structure the dependence of stream function maximum modulus in the first fluid on Mr number (line 1) is presented. The structure similar to this one but with the opposite vortex rotation direction (structure b) in the case of a flat interface is unstable with respect to disturbances breaking the motion symmetry with respect to the vertical axis and can be realized only under artificial symmetry conditions introduced in numerical calculation (line 2). A nonlinear development of instability leads finally to stabilization of motion with structure a. When heating from below the steady state keeps stability.

Let us describe now the calculation results corresponding to the contact angle value $\gamma = 80°$. When heating from above at any small Mr values the motion with structure b arises (line 3, Fig. 5.11). The shaded part of line 3 corresponds to the range of Mr values in which the given motion is unstable and can be got only if symmetry conditions are imposed. The dependence $|\psi_1|(Mr)$ does not have peculiarities at values $Mr \simeq 2500$. At $Mr = 4100$ the motion becomes unstable; however, the development of instability leads not to the transition to the motion of

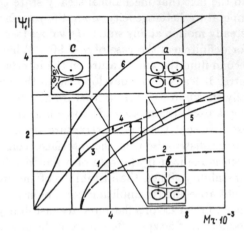

Figure 5.11. The amplitude curves and motion forms for a system with a deformable interface.

structure a, but to the statement of a non-symmetry motion with respect to the vertical axis (structure c, line 4, Fig. 5.11). This motion is realized only in the intermediate region of Mr values. At $Mr < 2400$ it goes on to the four-vortex motion of structure b, and at $Mr > 5700$ to the motion of structure a (line 5, Fig. 5.11). Stability regions of structures a and c in the interval $5100 < Mr < 5700$ overlap. In Fig. 5.12 a, b, for example, the pictures of isolines and isotherms, obtained in numerical calculation corresponding to the structure c of Fig. 5.11 are shown. Thus, a slight interface distortion may lead to the considerable changes of flow regime. In the case of heating from the side of the second fluid the temperature distribution causes the motion with structure a.

At large interface distortions the motion whose form is defined by the character of heating is realized. So, if the contact angle is $\gamma = 60°$, when heating from the side of the first fluid the flow with the structure b (line 6, Fig. 5.11) is realized and when heating from the side of the second fluid the motion with structure a is realized. The example of stream lines and isotherms pictures for $\gamma = 60°$ is presented in Fig. 5.13.

We shall discuss now the influence of an interface distortion on the oscillatory convection regimes. Let us consider the system with parameters $\eta = \nu = 0.5$; $\kappa = \chi = P = a = 1$; $L = 2.5$; heating is from the side of the second fluid. If $\gamma = 90°$, as a result of the steady state in-

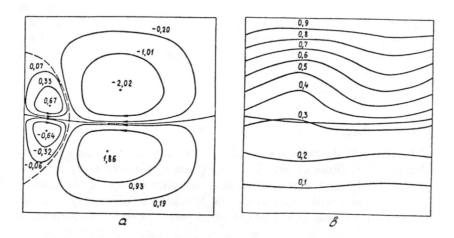

Figure 5.12. Isolines (a) and isotherms (b) for asymmetric flow (structure c Fig. 5.11); $Mr=3500$.

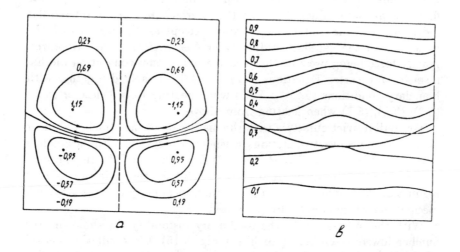

Figure 5.13. Isolines (a) and isotherms (b) for convective flow in the case of large interface deformation; $Mr=1000$, heating from above.

stability, the oscillatory convection regimes arise (see §5.1). In the case $\gamma \neq 90°$ at any small values of Mr number a steady convective motion is realized. As the calculations show, in the range $84° < \gamma < 90°$ the oscillatory instability of this motion leading to the development of oscillatory regimes takes place. For the description of oscillatory motions we shall use the variables $S_1(t)$ and $S_2(t)$ introduced by the formulas (4.20), and variables

$$S_{m+} = \max_t S_m(t), \quad S_{m-} = \min_t S_m(t), \quad m = 1, 2.$$

In Fig. 5.14 dependences $S_1(Mr)$ for steady motion (lines a), and also $S_{1+}(Mr)$ (lines b) and $S_{1-}(Mr)$ (lines c) for oscillatory motions relative to different values of angle γ are shown. As in the case of a flat interface, with the increase in Mr number oscillatory motions transit to steady ones by way of unlimited period growth. The oscillatory motions existence region decreases with the decrease in γ and disappears at $\gamma < 84°$.

5.3.   Finite-amplitude convection combining instabilities

Let us consider the effect of gravity on thermocapillary flows [109]. Note first that at finite values of gravity the interface may be considered flat independent of the contact angle values, except for the region with the dimension of the capillary radius order (see Section 1.2). Moreover, gravity force being present, the thermogravitational convection mechanism ($G \neq 0$) acts. Equations (4.1) — (4.6) are solved together with boundary conditions (5.2) — (5.5) when heating from below or (5.2), (5.3), (5.6), (5.7) when heating from above.

We shall restrict ourselves with the discussion of the Grashof number G influence on convective motions, arising in the system with parameters $\eta = \nu = 0.5; \kappa = \chi = \beta = P = a = 1; L = 2.5; \gamma = 90°$. In Fig. 5.15 the numerical experiments results carried out for the definition of flow regimes at the combined action of the thermocapillary and the Rayleigh mechanisms of instability are presented.

With the increase in G, the oscillatory instability threshold on Mr number lowers essentially (see line 1 Fig. 5.15); the oscillation period along the neutral stability line first decreases and then increases. Oscillatory instability leads to the soft generation of a periodic regime the space symmetry of which is defined by (5.10). The stream lines form essentially depends on the Grashof number G. At small G vortexes have a flattened to the interface form typical for a thermocapillary convec-

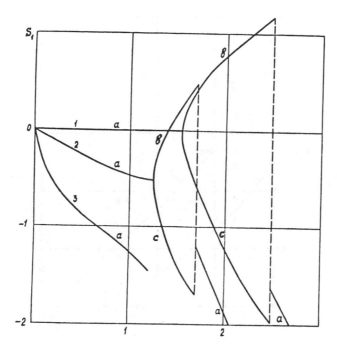

Figure 5.14. The amplitude curves for stationary and oscillatory convection
regimes relative to contact angle values 90° (line 1); 87° (2); 60° (3).

tion (the vortices centres are situated close to the interface); with the
increase in G the vortices centres recede from the interface. This im-
plies growing of the role of a thermogravitational (volume) instability
mechanism. The relevant patterns of a steady motion and temperature
distribution are shown in Fig. 5.16. In this case a convective circulation
fills all the cavity volume. The intensification of unharmonism and the
increase of oscillations period with the increase in Mr at fixed G values
are observed (see Fig. 5.17). At small Mr the marginal stability has a
monotonous character (see line 2 Fig. 5.15). This mode of instability is
of thermogravitational origin; the thermocapillary effect influences it in
a stabilizing manner (see Section 2.6).

If $G = 0$, the increase in Mr leads to transition from oscillatory to
steady motions. The transition picture, however, is essentially different for different G. At $G = 0$ the steady motion arose as a result of

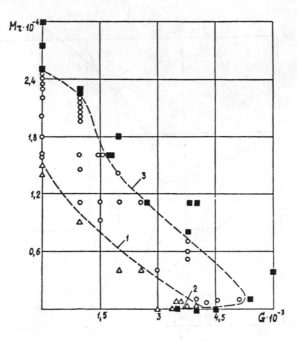

Figure 5.15. Composite map of convection regimes. Triangles relate to steady state, circles — to oscillatory regime, squares — to steady motion.

oscillations period turning to infinity; this motion possessed symmetry (5.10), and in region $0 < x < L/2$ for it $\psi_1 < 0$. At $G = 1000$ and $Mr \simeq 21500$ regular symmetric oscillations with the increase in Mr lose stability with respect to disturbances, breaking conditions (5.10). As a result of instability development, non-periodic oscillations arise (see Fig. 5.18). Further increase in Mr number leads to the establishment of a steady symmetric motion for which, however, $\psi_1 > 0$ as $0 < x < L/2$.

Another opportunity of transition between oscillatory and steady motions is realized at larger G values. So, as $Mr = 16000$ the existence regions of both motions types overlap in interval $1650 < G < 1750$. In Fig. 5.15 only the boundary of the region of oscillatory regimes existence is shown (line 3).

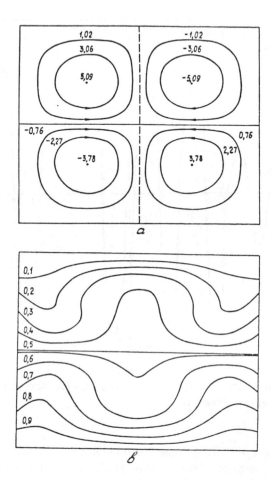

Figure 5.16. Isolines (a) and isotherms (b) of convective flow with the combination of thermocapillary effect and buoyancy force; Mr=4000; G=6000.

5.4. Convection with a temperature gradient along the interface

Let us pass on to the investigation of the situation when the temperature gradient applied to the system is directed along the interface. It is known that in this case at any weak temperature inhomogeneity along the interface the thermocapillary convection arises in the system. For an

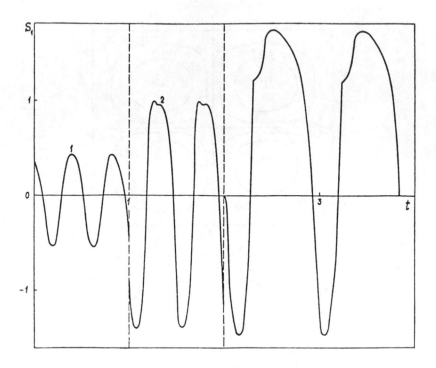

Figure 5.17. Oscillations form at fixed value G=1000; Mr=11000 (line 1); 16000 (2); 20000 (3).

infinite fluid layer with a nondeformable flat free surface the convection equations at constant temperature gradient admit the exact solution, describing a plane-parallel flow; this solution was first reported in [16]. Generalization of the solution for a two-layer fluid is realized in [128]. Let us pay attention to the fact that for a plane -parallel flow the pressure gradients along the layers are not equal to zero and differ from each other, therefore this free surface or interface stays flat only within the limit of the infinite surface tension. Under real conditions for extensive layers the distortion of an interface becomes essential [195, 133]; it can even lead to break up of the layer.

At arbitrary heterogeneous temperature distribution on the solid boundary, the analytical calculation of a thermocapillary flow is possible in a creep-flow approach, i.e. in the neglection by nonlinear terms in the motion equations and in the equation of heat transfer. An example of a

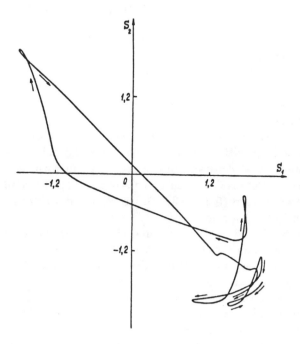

Figure 5.18. The section of phase trajectory for non-periodical oscillations; $Mr=21500$.

similar calculation are the results obtanied in [11]. This paper deals to the investigation of a thermocapillary convection for a series of systems: for a spherical fluid drop, a system of two immiscible fluids in a spherical vessel, a spherical fluid layer round a solid nucleus. The convection was caused by a heterogeneous temperature distribution on the external sphere surface. A nonlinear character of equations allowed to get a common problem solution in the series form. We shall also mention [27] in which the calculation of thermocapillary flow in a layer with free surface partially covered by a surfactant is carried out. In creep-flow approach an analytical solution is obtained describing the vortex system whose amplitude exponentially decreases deep into the stagnation region.

In a nonlinear statement thermocapillary flows induced by longitudinal temperature gradient were actively studied in connection with the analysis of noncrucible zone melting and directional cristallization process [77, 76, 172, 179]. In particular, the transitions between different convection regimes in a cylindrical fluid zone were studied experimentally [154, 155, 189]. The consecutive numerical analysis of an axissymmetric thermocapillary flow in a liquid zone (the distortion of a free surface being taken in consideration) is performed in [168]. Practically all the investigations in this area have been carried out in a one-layer approach. We shall mention, however, the paper [100] in which an attempt of an experimental investigation of convection in a liquid zone, keeping the interface betwen two media was undertaken.

As the simplest example of a nonlinear calculation in a two- layer approach we shall consider thermocapillary convection in a rectangular cavity, filled up with two immiscible fluids; lateral boundaries of the cavity are kept at constant and nonequal temperatures (see 1 Section 4.5). The problem is described by equations (5.1); the conditions on the interface keep form (5.2)–(5.3). On solid boundaries conditions (4.21), (4.22) and (4.24) are imposed.

In Fig. 5.19 and 5.20 the calculated results of thermocapillary flow realized in a system with parameter values $\eta = \nu = 0.5$, $\kappa = \chi = \beta = P = 1$ as $a = 1$ are shown [105]. The points at which the stream function maxima are reached with the increase in Mr number are shifted towards the interface and to the right. Similar flow calculations under a combined action of thermocapillary and concentration-capillary convection mechanisms (in one-layer approach) are described in the book [127]. We shall also note the article [99] in which in one-layer statement the generation of thermocapillary convection at a sudden temperature increase on one of the lateral walls is numerically investigated.

Paper [132] concerns thermocapillary and mixed convection (induced by combined action of thermogravitational and thermocapillary mechanisms) in a gap filled with two media circle between concentric cylinders on which constant and different temperature is given. Calculations were carried out for an air–water system. Two series of calculations are presented: 1) at a fixed Rayleigh number value, the Marangoni number varied; 2) at a fixed Marangoni number value, the Rayleigh number varied. Stream lines, isotherms and also local Nusselt number distributions along the boundaries of internal and external cylinders are obtained.

5.5.   Thermocapillary drift of a drop

Let us now pass on to the analysis of a situation, when one of the media forms a drop or a bubble in the other media. When a heterogeneous heating, the tangential stresses, stipulated by thermocapillary effect, may cause a drop motion as a whole. This phenomenon, called thermo-

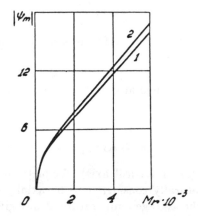

Figure 5.19. Dependence of stream function maximum value $|\psi_m|$ on Mr number in the first (line 1) and the second (line 2) fluids.

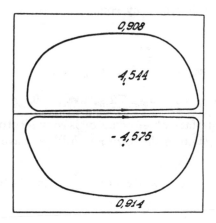

Figure 5.20. Stream lines picture for the model system as $Mr=750$.

capillary drift, was investigated for the first time in [192]. References can be found in [19].

Let a drop of media 2 be situated inside any enclosed surface Γ. In media 1 on a large distance from the drop a constant vertical temperature gradient exists:

$$|r| \to \infty : \frac{dT_1}{dz} = A. \tag{5.23}$$

We assume that the drop performs a uniform motion in a vertical direction with any velocity u, unchanged axis-symmetric interface form being kept. It is convinient to pass on to the reference system in which the drop is motionless; then at infinity the velocity of the first fluid has a constant value:

$$|r| \to \infty : v_1 = -u\gamma \tag{5.24}$$

(γ is the unit vector of vertical axis). Velocity, pressure and temperature fields are naturally considered to be axially -symmetric. In our reference system the average temperature at any given point changes at a constant rate:

$$\frac{\partial T}{\partial t} = uA.$$

The problem is described by equations (notation is the same as in Section 1.2):

$$(v_m \nabla)v_m = -\frac{1}{\rho_m}\nabla p_m + \nu_m \nabla v_m,$$

$$uA + v_m \nabla T_m = \chi_m \nabla T_m, \mathrm{div}\, v_m = 0; \quad m = 1, 2. \tag{5.25}$$

On a steady interface whose equation is conviniently written in spherical coordinate system in the form $r = l(\theta)$, we have:

$$(p_1 - \rho_1 gl\cos\theta) - (p_2 - \rho_2 gl\cos\theta) + \frac{\sigma}{R} = (\sigma_{1,ik} - \sigma_{2,ik})n_i n_k;$$

$$(\sigma_{1,ik} - \sigma_{2,ik})\tau_i n_k - \alpha\tau_i\frac{\partial T_1}{\partial x_i} = 0;$$

$$v_{1,i}n_i = v_{2,i}n_i = 0; \quad (v_{1,i} - v_{2,i})\tau_i = 0; \tag{5.26}$$

$$T_1 = T_2; \quad \left(\kappa_1\frac{\partial T_1}{\partial x_i} - \kappa_2\frac{\partial T_2}{\partial x_i}\right)n_i = 0.$$

Let us pass on to dimensionless variables, taking for length, time, velocity, pressure and temperature units the values a, a^2/ν_1, ν_1/a, $\rho_1\nu_1^2/a^2$ and Aa where a is the average radius of the drop.

Equations and boundary conditions take the form:

$$(v_1\nabla)v_1 = -\nabla p_1 + \Delta v_1,$$

$$Re + v_1\nabla T_1 = \frac{1}{P}\Delta T_1, \quad \text{div}\, v_1 = 0; \tag{5.27}$$

$$(v_2\nabla)v_2 = -\rho\nabla p_2 + \frac{1}{\nu}\Delta v_2,$$

$$Re + v_2\nabla T_2 = \frac{1}{\chi P}\Delta T_2, \quad \text{div}\, v_2 = 0, \tag{5.28}$$

$$r \to \infty : \frac{dT_1}{dz} = 1, \quad v_1 = -Re\gamma;$$

$r = l(\theta):$

$$p_1 - p_2 + \frac{W_o - MT}{\mathcal{R}} + Ga(\rho^{-1} - 1)l\cos\theta =$$

$$\left[\left(\frac{\partial v_{1,i}}{\partial x_k} + \frac{\partial v_{1,k}}{\partial x_i}\right) - \eta^{-1}\left(\frac{\partial v_{2,i}}{\partial x_k} + \frac{\partial v_{2,k}}{\partial x_i}\right)\right]n_i n_k;$$

$$\left[\left(\frac{\partial v_{1,i}}{\partial x_k} + \frac{\partial v_{1,k}}{\partial x_i}\right) - \eta^{-1}\left(\frac{\partial v_{2,i}}{\partial x_k} + \frac{\partial v_{2,k}}{\partial x_i}\right)\right]\tau_i n_k - \tag{5.29}$$

$$- M\tau_i\frac{\partial T_1}{\partial x_i} = 0;$$

$$v_{1,i}n_i = v_{2,i}n_i = 0; \quad (v_{1,i} - v_{2,i})\tau_i = 0;$$

$$T_1 = T_2; \quad \left(\kappa\frac{\partial T_1}{\partial x_i} - \frac{\partial T_2}{\partial x_i}\right)n_i = 0.$$

where $Re = ua/\nu_1$ is the Reynolds number, $Ga = ga_1^3/\nu_1^2$ is the Galileo number, $M = \alpha A a^2/\eta_1\nu_1$ is similar to the Marangoni number, $W_0 = \sigma_0 a/\eta_1\nu_1$; all the other parameters are the same as in Section 1.2.

Since M is small, the problem solution (5.27)–(5.29) can be constructed on the basis of expansion in parameter M:

$$v_i = v_i^{(0)} + v_i^{(1)}M^2 v_i^{(2)} + \cdots,$$

$$p_i = p_i^{(0)} + M p_i^{(1)} + M^2 p_i^{(2)} + \cdots,$$

$$T_i = T_i^{(0)} + M T_i^{(1)} + M^2 T_i^{(2)} + \cdots,$$

$$Re = Re^{(0)} + M Re^{(1)} + M^2 Re^{(2)} + \cdots,$$

$$l(\theta) = 1 + M l^{(1)}(\theta) + M^2 l^{(2)}(\theta) + \cdots$$

In the lowest order in M calculations of the flow and of variable Re in the limit $Re \ll 1$ were done in [192] (see also [190]); the formula defining the drop velocity drift is:

$$Re = \frac{2Ga(1 - \rho^{-1}(1 + \eta^{-1}))}{3(2 + 3\eta^{-1})} + \frac{2M}{(2 + \kappa^{-1})(2 + 3\eta^{-1})}.$$

Calculation of drop motion velocity in the opposite limit case ($Re \gg 1$) is performed in [130]. Under assumption that the Pekle number is $Pe = ReP \ll 1$ (convective heat exchange is not essential) and the Weber number is $We = Re^2/W_0 < 2$ (therefore the drop keeps a spherical form), the expression is:

$$Re = \frac{2Ga(1 - \rho^{-1})}{9(2 + 3\eta^{-1})} + \frac{2M}{(2 + \kappa^{-1})(2 + 3\eta^{-1})}.$$

At $Pe \gg 1$ and $We < 2$ in the limit $\kappa, \eta \to \infty$ we find:

$$Re = \frac{Ga(1 - \rho^{-1})}{9} + \frac{M}{3}.$$

Calculations in higher orders for microgravity ($Ga = 0$) were carried out by Yu.K. Bratukhin [17, 18]. Nonsteady drop drift at the speed - up stage was investigated in [5, 130]. We shall also point to the article [82], in which a related problem on viscous fluid drop motion in the solution of surfactant with constant concentration is considered.

Thermocapillary drift of a drop may also take place in a medium with constant temperature if the system possesses internal sources of temperature heterogeneity. For example, in [65, 148] drop motion with nonisothermal chemical reaction proceeding on its surface with any heat effect Q is considered. It is found that in case $Qd\sigma/dT < 0$ the drift of a drop with any constant velocity (the drift direction is arbitrary) is steady. Drop motion under the conditions of homogeneous internal heat release or homogeneous heat release on the surface was studied in [137].

At finite values of parameter M the solution may be obtained only numerically. The flow calculation algorithms including a drop form deviation from spherical are constructed by V.Ya. Rivkind with co-authors [143–146]. At $W_0 \gg 1$ these deviations are not considerable and may be neglected.

As stated in [146], with the increase in M the flow picture in a drop and in the surrounding fluid changes. With the increase in M in stagnation region behind the drop a wake vortex appears, which induces inside the drop a vortex with the sign opposite to the initial one. With further increase in M the wake vortex in the first fluid covers the drop and the direction of the fluid motion in a drop changes to the opposite.

A qualitative experimental investigation of a thermocapillary drift was carried out for the first time in [192]. In [20] measurement methods of a bubble thermo-drift in a vertical section heated from the side close to the inverse point of heat expansion coefficient were devised. Quantitative measurements of a thermocapillary drift velocity based on these methods are presented in [19].

CHAPTER 6

Method of Markers and Cells for Studying Convection Regimes

Calculations of the finite-amplitude convection regimes in Chapters 5 and 6 were carried out under assumption of a nondeformable interface (infinitely large surface tension). This assumption may be not fulfilled, in particular, if the surface tension on the interface is small and the fluids have close densities or more heavy fluid is situated above (see § 2.7). In this case convection generation may lead to the deformation of the interface and even to mixing of fluids. In this chapter attempts at studying the convection regimes in a two-layer system with a deformable interface heated from below or from above are described. Combined influence of the convective and the Rayleigh–Taylor mechanisms of instability is investigated; thermocapillary effect is not taken into consideration. To study convective mixing ordinary finite-difference methods using the statement of boundary conditions on the interface are inefficient. Actually, because of mutual fluids penetration this surface takes an extremely complicated form. In order to overcome these difficulties it is possible to use the two - fluids variant of marker and cell method [129, 34, 35]. The scheme of method is described in Section 6.1. Application of the method to the calculation of free convection in a two-layer system is considered in Section 6.2. Calculation results for the problem of the convection generation in a two-layer system with internal heat sources are described in the same section.

6.1. General scheme of method of markers and cells

The method of markers and cells [70, 71] operates with Lagrangian particles (markers) moving against a background of Euler calculation mesh. In the two-fluid variant of the method, markers of two types are used, marking the particles of the first and the second fluids. Each marker moves in accordance with the magnitude and direction of velocity at the given place. The velocity field is calculated on the basis of equations determining the motion of a homogeneous medium whose characteristics (density, viscosity and so on) change from point to point and in time. The change of these characteristics, stipulated by mixing of fluids, is defined by the displacements of the markers. Thus, the markers play a dual role in the scheme. In the first place they show the space regions occupied by each fluid and also the relative fluids content in the mixing zones; in the second place they define the variables of density, viscosity and other charecteristics necessary for calculation of markers' displacement.

Density in any cell of calculating mesh is defined by

$$\rho = (\rho_1 N_1 + \rho_2 N_2)(N_1 + N_2)^{-1}, \qquad (6.1)$$

where ρ_1 and ρ_2 are respective fluid densities, and N_1 and N_2 are the markers' numbers type in the mesh cell; other fluids characteristics are calculated by similar formulas.

Let us consider a cavity of rectangular cross section $0 \leq x \leq l$, $-a_2 \leq z \leq a_1$, filled up with two viscous incompressible fluids; at the reference state the interface is horizontal: $z = 0$. All the cavity boundaries are solid, horizontal boundaries $z = -a_2$ and $z = a_1$ are kept at constant different temperatures $T = \theta$ and 0 (case $\theta > 0$ corresponds to heating from below and $\theta < 0$ — to heating from above); along the vertical boundaries $x = 0$ and $x = l$ the temperature distribution providing an opportunity of mechanical steady state is assumed.

Equations defining a flat convective motion of the media, dynamic and thermodynamic characteristics of which depend on coordinates, have the form

$$\frac{\partial(\rho u)}{\partial t} + \frac{\partial(\rho u^2)}{\partial x} + \frac{\partial(\rho u v)}{\partial z} =$$
$$-\frac{\partial p}{\partial x} + 2\frac{\partial}{\partial x}\left(\eta\frac{\partial u}{\partial x}\right) + \frac{\partial}{\partial z}\left[\eta\left(\frac{\partial u}{\partial z} + \frac{\partial v}{\partial x}\right)\right], \quad (6.2)$$

$$\frac{\partial(\rho v)}{\partial t} + \frac{\partial(\rho u v)}{\partial x} + \frac{\partial(\rho v^2)}{\partial z} =$$
$$-\frac{\partial p}{\partial z} + \frac{\partial}{\partial x}\left[\eta\left(\frac{\partial u}{\partial z} + \frac{\partial v}{\partial x}\right)\right] + 2\frac{\partial}{\partial z}\left(\eta\frac{\partial v}{\partial z}\right) - \rho g(1 - \beta T), \quad (6.3)$$

$$\rho C_p \left(\frac{\partial T}{\partial t} + u\frac{\partial T}{\partial x} + v\frac{\partial T}{\partial z}\right) = \frac{\partial}{\partial x}\left(\kappa\frac{\partial T}{\partial x}\right) + \frac{\partial}{\partial z}\left(\kappa\frac{\partial T}{\partial x}\right), \quad (6.4)$$

$$\frac{\partial u}{\partial x} + \frac{\partial v}{\partial z} = 0. \qquad (6.5)$$

Here u and v are velocity components along the axis x, and z, p is pressure, T is temrerature, g is the gravitational acceleration, η, κ, β and

C_p are coefficients of dynamic viscosity, heat conductivity, heat expansion and specific heat, being at the same time functions of coordinates owing to mutual fluids penetration.

It is necessary to explain that in the situation under consideration the conservative law of mass mixing (media with variable density) is satisfied if in parallel to incompressibility condition (6.5) the condition of Lagrangian derivative of density on time vanishing is satisfied

$$\frac{d\rho}{dt} = 0.$$

This condition is not used in that part of calculation where the velocity and pressure fields are determined, but it is used for determination of the density field at the next time step; it puts restrictions on the possible markers' motions (see lower).

Let us go to dimensionless variables in equations (6.2)–(6.5) using for units of length, time, velocity, pressure and temperature the values of $a_1, \rho_1 a_1^2/\eta_1, \eta_1/rho_1 a_1, \eta_1^2/a_1^2\rho_1$ and θ:

$$\frac{\partial(\tilde{\rho}u)}{\partial t} + \frac{\partial(\tilde{\rho}u^2)}{\partial x} + \frac{\partial(\tilde{\rho}uv)}{\partial z} =$$
$$-\frac{\partial p}{\partial x} + 2\frac{\partial}{\partial x}\left(\tilde{\eta}\frac{\partial u}{\partial x}\right) + \frac{\partial}{\partial z}\left[\tilde{\eta}\left(\frac{\partial u}{\partial z} + \frac{\partial v}{\partial x}\right)\right], \quad (6.6)$$

$$\frac{\partial(\tilde{\rho}v)}{\partial t} + \frac{\partial(\tilde{\rho}uv)}{\partial x} + \frac{\partial(\tilde{\rho}v^2)}{\partial z} =$$
$$-\frac{\partial p}{\partial z} + \frac{\partial}{\partial x}\left[\tilde{\eta}\left(\frac{\partial u}{\partial z} + \frac{\partial v}{\partial x}\right)\right] + 2\frac{\partial}{\partial z}\left(\tilde{\eta}\frac{\partial v}{\partial z}\right) + G\tilde{\rho}\tilde{\beta}T - \tilde{\rho}Ga, \quad (6.7)$$

$$\frac{\partial T}{\partial t} + u\frac{\partial T}{\partial x} + v\frac{\partial T}{\partial z} = \frac{1}{P\tilde{\rho}\tilde{C}_p}\left[\frac{\partial}{\partial x}\left(\tilde{\kappa}\frac{\partial T}{\partial x}\right) + \frac{\partial}{\partial z}\left(\tilde{\kappa}\frac{\partial T}{\partial z}\right)\right], (6.8)$$

$$\frac{\partial u}{\partial x} + \frac{\partial v}{\partial z} = 0. \quad (6.9)$$

The variable values $\tilde{\rho}, \tilde{\eta}, \tilde{\kappa}, \tilde{\beta}, \tilde{C}_p$ which enter the equations are also dimensionless and are defined as ratios of media parameters to fixed values $\rho_1, \eta_1, \kappa_1, \beta_1, C_{p_1}$. In this section , the sign " $\sim$ " will be omitted. As before, characteristics of the fluid filling up the upper half of the cavity at the steady state are marked by index 1. The Galileo number Ga, the Grashof number G and the Prandtl number P are defined by

$$Ga = \frac{\rho_1^2 g a_1^3}{\eta_1^2}, \quad G = \frac{g\beta_1\theta a_1^3\rho_1^2}{\eta_1^2}, \quad P = \frac{\eta_1 C_{p_1}}{\kappa_1}.$$

It is necessary to note that in (6.6) and (6.7) the variable p indicates the full pressure (rather than the pressure counted from hydrostatic one). On the solid walls the velocity vanishes:

$$u = v = 0.$$

On the horizontal cavity boundaries (when heating from below):

$$z = -a: \quad T = 1,$$
$$z = 1: \quad T = 0.$$

The temperature distribution along the vertical boundaries has the form

$$T = \frac{1-z}{1+\kappa a} \quad (z \geq 0),$$
$$T = \frac{1-\kappa z}{1+\kappa a} \quad (z \leq 0).$$

Boundary conditions have a similar form when heating is from above (see formulas (5.7)).

For numerical solution of equations (6.6)–(6.9) the divergence difference scheme in which the conservation laws are fulfilled for each cell of rectangular calculation mesh, is used. Divergence difference correlations being put down the horizontal and vertical velocity components are localized in the centers of vertical and horizontal cell walls, and all the scalar variables are related to the cells' centres (Fig. 6.1). Let us write down the finite-difference analog of equations (6.6)–(6.9). Lower i and j number everywhere vertical and horizontal coordinate lines, passing through the cells centres (see Fig. 6.1), n is the time circle number and Δt is the time step.

$$(\rho u)_{i+1/2,j}^{n+1} = (\rho u)_{i+1/2,j}^{n} + \Delta t \left(Q_{i+1/2,j}^{n} + \frac{p_{i,j}^{n} - p_{i+1,j}^{n}}{\Delta x} \right), \qquad (6.10)$$

$$(\rho v)_{i,j+1/2}^{n+1} = (\rho v)_{i,j+1/2}^{n} + \Delta t \left(S_{i,j+1/2}^{n} + \frac{p_{i,j}^{n} - p_{i,j+1}^{n}}{\Delta z} \right), \qquad (6.11)$$

$$T_{i,j}^{n+1} = T_{i,j}^{n} + \Delta t E_{i,j}^{n}, \qquad (6.12)$$

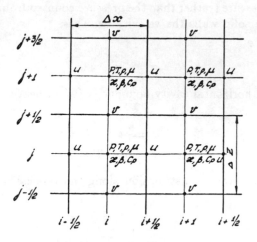

Figure 6.1. Calculation net.

$$\frac{u^{n+1}_{i+1/2,j} - u^{n+1}_{i-1/2,j}}{\Delta x} + \frac{v^{n+1}_{i,j+1/2} - v^{n+1}_{i,j-1/2}}{\Delta z} = 0. \qquad (6.13)$$

Here the following notation is introduced:

$$Q_{i+1/2,j} = \frac{(\rho u^2)_{i,j} - (\rho u^2)_{i+1,j}}{\Delta x}$$
$$+ \frac{(\rho u v)_{i+1/2,j-1/2} - (\rho u v)_{i+1/2,j+1/2}}{\Delta z}$$
$$+ 2\frac{\eta_{i+1,j}(u_{i+3/2,j} - u_{i+1/2,j}) - \eta_{i,j}(u_{i+1/2,j} - u_{i-1/2,j})}{(\Delta x)^2}$$
$$+ \frac{\eta_{i+1/2,j+1/2}[(u_{i+1/2,j+1} - u_{i+1/2,j})/\Delta z + (v_{i+1,j+1/2} - v_{i,j+1/2})/\Delta x]}{\Delta z}$$
$$- \frac{\eta_{i+1/2,j-1/2}[(u_{i+1/2,j} - u_{i+1/2,j-1})/\Delta z + (v_{i+1,j-1/2} - v_{i,j-1/2})/\Delta x]}{\Delta z}$$

$$S_{i,j+1/2} = \frac{(\rho v^2)_{i,j} - (\rho v^2)_{i,j+1}}{\Delta z} + \frac{(\rho uv)_{i-1/2,j+1/2} - (\rho uv)_{i+1/2,j+1/2}}{\Delta x}$$

$$+ 2\frac{\eta_{i,j+1}(v_{i,j+3/2} - v_{i,j+1/2}) - \eta_{i,j}(v_{i,j+1/2} - v_{i,j-1/2})}{(\Delta z)^2}$$

$$+ \frac{\eta_{i+1/2,j+1/2}[(u_{i+1/2,j+1} - u_{i+1/2,j})/\Delta z + (v_{i+1,j+1/2} - v_{i,j+1/2})/\Delta x]}{\Delta x}$$

$$- \frac{\eta_{i-1/2,j+1/2}[(u_{i-1/2,j+1} - u_{i-1/2,j}/\Delta z + (v_{i,j+1/2} - v_{i-1,j+1/2})/\Delta x]}{\Delta x}$$

$$+ G(\rho\beta T)_{i,j+1/2} - p_{i,j+1/2} \cdot Ga;$$

$$E_{i,j} = u_{i,j}\frac{T_{i-1/2,j} - T_{i+1/2,j}}{\Delta x} + v_{i,j}\frac{T_{i,j-1/2} - T_{i,j+1/2}}{\Delta z}$$

$$+ \frac{1}{P\rho_{i,j}C_{p_{i,j}}}\left\{\left[\frac{\kappa_{i+1/2,j}(T_{i+1,j} - T_{i,j}) - \kappa_{i-1/2,j}(T_{i,j} - T_{i-1,j})}{(\Delta x)^2}\right]\right.$$

$$+ \left.\left[\frac{\kappa_{i,j+1/2}(T_{i,j+1} - T_{i,j}) - \kappa_{i,j-1/2}(T_{i,j} - T_{i,j-1})}{(\Delta z)^2}\right]\right\}.$$

Some variables in (6.10)–(6.13) are not defined directly by the fields of u, v, ρ and η are calculated by interpolation. For example,

$$u_{i,j} = \frac{1}{2}(u_{i+1/2,j} + u_{i-1/2,j}), \tag{6.14}$$

$$v_{i,j} = \frac{1}{2}(v_{i,j+1/2} + v_{i,j-1/2}), \tag{6.15}$$

$$p_{i+1/2,j+1/2} = \frac{1}{4}(\rho_{i+1,j} + \rho_{i,j} + \rho_{i,j+1} + \rho_{i+1,j+1}). \tag{6.16}$$

Substituting values of u_{n+1} and v_{n+1} calculated from (6.10) and (6.11) into equation (6.13) it is possible to get the equation of Poisson type for pressure. Let us write the formula for the solution of this equation by iterative method (the Libman scheme with a consequetive upper relaxation); in this formula s is the iteration number and ω is the relaxation parameter ($\omega = 1.5$)

$$p_{i,j}^{s+1} = p_{i,j}^s + \omega(p_{i,j}^{s+1/2} - p_{i,j}^s), \tag{6.17}$$

$$p_{i,j}^{s+1/2} = \left[\frac{1}{(\Delta x)^2} \left(\frac{p_{i+1,j}}{\rho_{i+1/2,j}} + \frac{p_{i-1,j}}{\rho_{i-1/2,j}} \right) \right.$$

$$+ \frac{1}{(\Delta z)^2} \left(\frac{p_{i,j+1}}{\rho_{i,j+1/2}} + \frac{p_{i,j-1}}{\rho_{i,j-1/2}} \right) - R_{i,j} \right] \bigg/ \left[\frac{1}{(\Delta x)^2} \left(\frac{1}{\rho_{i+1/2,j}} + \frac{1}{\rho_{i-1/2,j}} \right) \right.$$

$$+ \frac{1}{(\Delta z)^2} \left(\frac{1}{\rho_{i,j+1/2}} + \frac{1}{\rho_{i,j-1/2}} \right) \bigg],$$

$$\tag{6.18}$$

where

$$R_{i,j} = \frac{1}{\Delta t} \left(\frac{u_{i+1/2,j} - u_{i-1/2,j}}{\Delta x} + \frac{v_{i,j+1/2} - v_{i,j-1/2}}{\Delta z} \right) +$$

$$\frac{1}{\Delta x} \left(\frac{Q_{i+1/2,j}}{\rho_{i+1/2,j}} - \frac{Q_{i-1/2,j}}{\rho_{i-1/2,j}} \right) + \frac{1}{\Delta z} \left(\frac{S_{i,j+1/2}}{\rho_{i,j+1/2}} - \frac{S_{i,j-1/2}}{\rho_{i,j-1/2}} \right).$$

It is convenient to formulate the boundary conditions for the difference equations introducing the additional (fictitious) cell rows situated behind the boundaries of the calculated region. By choosing function values in these cells in the appropriate manner, it is possible to fulfil the boundary conditions and simultaneously to standardize calculations at all points (including also those close to the boundary).

Let us consider the relevant formulas for solid impenetrable boundary perpendicular to z axis. On such a boundary both velocity components and the density derivative with respect to normal to the boundary must vanish

$$u = v = 0, \quad \frac{\partial \rho}{\partial z} = 0. \tag{6.19}$$

Condition $u = 0$ and a formula of type (6.14) lead to

$$u_{i+1/2,0} = -u_{i+1/2,1}. \tag{6.20}$$

For obtaining conditions for v, it is necessary, along with (6.19), to use continuity equation (6.9) which leads to the relation

$$v_{i,1/2} = 0, \quad v_{i,-1/2} = v_{i,3/2}. \tag{6.21}$$

For density from (6.19) we get

$$p_{i,0} = p_{i,1}.$$ (6.22)

The values of η at additional points calculated from similar relations

$$\eta_{i,0} = \eta_{i,1}.$$ (6.23)

Using (6.20)–(6.23) and (6.11), it is possible to find the pressure values at fictitious points

$$p_{i,0} = p_{i,1} - 4\eta_{i,1} \cdot v_{i,3/2}/\Delta z - [(\eta_{i,1} + \eta_{i+1,1}) \times$$
$$u_{i+1/2,1} - (\eta_{i,1} + \eta_{i-1,1})u_{i-1/2,1}]/\Delta x.$$ (6.24)

The conditions on solid boundaries, orientated perpendicularly to the axis x are written in similar manner.

Definition of pressures, velocities and temperatures from differential equations at given distributions of $\rho, \eta, \kappa, \beta$ and C_p is carried out in the following order. First by the iteration method the pressure field is found. Calculation by (6.17)–(6.18) is performed until the relative error in each point does not become smaller than a preset small variable. Then velocity fields are calculated from equations (6.10) and (6.11) in which the obtained pressure field is used again, the temperature field is calculated from equation (6.12), and at last the variables values at points close to the boundaries are determined from (6.20)–(6.24).

Let us consider now the calculation of densities and viscosities fields upon transition to the next time step. New density and viscosity fields are found as a result of iteration process, the aim of which is to satisfy condition $d\rho/dt = 0$. Each step of this process includes the above calculation complex for determination of velocities, temperatures and pressures, calculation of new markers' coordinates

$$x = x^n + \Delta t u^{n+1}, \quad z = z^n + \Delta t v^{n+1}$$ (6.25)

(markers' velocities are found by the interpolation) and determination of density, viscosity and coefficients β, κ and C_p values in each cell by formulas similar to (6.1). Again, the obtained density values are compared with ρ values which have been used in calculation of pressures and velocities. If these values do not coincide within the desired accuracy, then another iteration step is done and the markers return to their former places, but in the calculation of velocities, temperatures and pressures again the obtained density and viscosity fields are used. The process being over (coincidence of the calculated density field with the initial one) the calculation cycle in the limits of the current time step is considered to be finished. The time step value is selected from the calculation stabilization conditions in the following manner:

$$\Delta t = \min \frac{(\Delta x)^2 (\Delta z_m)^2}{4[(\Delta x)^2 + (\Delta z_m)^2]}, \quad m = 1, 2.$$

6.2.   Convection regimes with a deformable interface

1. Convection when heating from below. Let us pass on to the description of the results of [165, 166]. All calculations are carried out for the case $L = 2, a = 1$. For the initial state the mechanical reference state is used in which the interface is situated in the middle of the cavity. In this case the velocity in all points is zero, temperature T depends only on coordinate z, and pressure field is determined from the equation

$$\frac{\partial \rho}{\partial z} = G \tilde{\rho} \tilde{\beta} T - \tilde{\rho} Ga.$$

Disturbances imposed on the initial state at the initial time moment do not involve velocity and pressure fields. The disturbances have the form of temperature deviations from individual non-symmetrically distributed points.

Numerical model of motions arising in a two-layer system was made for fluids close in physical properties. As the calculations show, the marginal stability and motions structure at the given degree of heating depend mainly on the fluids densities difference (and in much smaller degree on the difference in other parameters). That is why the basic calculations relate to the case when the values of all parameters (except density) are the same for both fluids; the values of Prandtl and Galileo numbers are fixed: $P = 1$ and $Ga = 10^5$. The Galileo number was chosen so as to use the Boussinesq approach (see Section 1.2). The Boussinesq approach is justified if density increments stipulated by heat expansion are small in comparison with average density value. This requirement leads to the condition $\delta_\beta = \beta_1 |\theta| \ll 1$ which puts restriction on the values of the Galileo number and the Grashof number: $\delta_\beta = G/Ga \ll 1$. As $Ga = 10^5$ and G change in the limits from 0 to 10^4, the Boussinesq approach may be considered applicable. Such selection of variables leads to the fact that the problem is completely defined by two main parameters: the Grashof number G and the densities ratio $\rho = \rho_1/\rho_2$. Values of G and ρ being used, the development of initial disturbances led to one of the following regimes: mechanical reference state, Rayleigh–

Taylor instability, a two-layer convective structure with practically flat horizontal interface and global circulation of the fluid in all the cavity with the interface destruction. A summary diagram of regimes realized in a two-layer system is presented in Fig. 6.2. The region $G > 0$ corresponds to heating from below, and $G < 0$ — to heating from above. The reference states stabilized as a result of disturbances development at the given parameters G and ρ are marked by triangles, the states at which the system keeps a two-layer structure are marked by circles, and the states at which the interface is destroyed and the fluids mix are marked by squares. In cases when the mixing process proceeds quickly it was treated as development of Rayleigh– Taylor instability (squares in Fig. 6.2). If the typical time of mixing development is considerable, it is natural to assume that it is stipulated by convective mechanism (circles in Fig. 6.2 correspond to this case).

Let us discuss first the results corresponding to the case $G = 0$ (heating is absent). It is known that in an isothermal situation the reference state of a two-layer system may become unstable if the heavy fluid is situated above the light one (Rayleigh–Taylor instability). The surface tension being of any small value, Rayleigh–Taylor instability must arise if the upper fluid density is on any small value larger than the lower fluid density. Really, in calculations at the values $G = 0$ and $\rho = 0.999$ the down fall of the upper fluid into the lower one is observed, and in the calculation for $G = 0$ and $\rho = 1$ the system is kept at the steady state.

Passing on to the results for $G \neq 0$, we shall note that when heating from above ($G < 0$), the Rayleigh- Taylor mechanism is the only reason for instability in the system. The calculations corresponding to the region $G < 0$ demonstrate the existence of an interesting effect — heating from above shifts the threshold of Rayleigh–Taylor instability in the range $\rho < 1$. For example, at $G = -4000$ (see Fig. 6.2) the steady state is stable also at $\rho = 0.999$.

In the range $G > 0$ (heating from below) besides the Rayleigh–Taylor mechanism the convective mechanism of instability acts. The steady state of a homogeneous fluid filling up a square cavity is known to become unstable [48] when the Grashof number exceeds $G_* \simeq 640$ (in the accepted units), and the motion breaking the stability is of one-vortex structure. Calculations for $\rho = 1$ (a homogeneous fluid) give the value of the critical Grashof number G_* close to that known from the linear theory. It can serve as a control of the method used.

In a two-layer system the difference of fluids' densities (at $\rho > 1$) can shift the stability threshold in the region $G > G_*$. It is probably connected with the fact that the density shock on the interface prevents

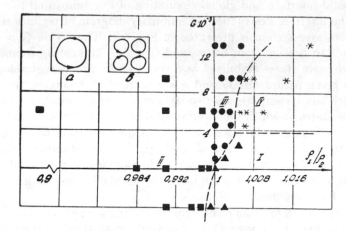

Figure 6.2. Composite regimes diagram. Regions of mechanical steady state (1), Rayleigh–Taylor instability (2) convective mixing (3), two-layer convective structure (4).

the development of the motion connected with the fluid flow through the interface.

Density shock can stabilize the steady state of a two-layer system only at $G \leq 3800$; at $G > 3800$ the stability is broken by a four-vortex motion the existence of which is not connected with the fluid flow through the interface [166]. At $G > 3800$ depending on the densities ratio value two convection regimes are possible: one-vortex flow mixing both fluids, and a four-vortex motion. The latter is not connected with the flow through the interface and thus keeps a two-layer system structure (the form of the given motions stream lines is schematically shown in the inserts a and b to Fig. 6.2). The first convection regime exists at the densities ratio ρ close to unity; starting with any value of ρ this regime is not realized, and the regime of a four-vortex motion arises. Sufficient difference between densities seems to maintain the interface on the system, preventing the motions typical for a homogeneous fluid.

Fig. 6.3 and 6.4 illustrate the onset and the development in time of the two convection regimes described (Fig. 6.3 corresponds to $G = 6000$, $\rho = 1.005$, Fig. 6.4 corresponds to $G = 6000$, $\rho = 1.001$). In each photograph the markers position marking the particles of the first and the second fluids at any moment are shown. Markers of the fluid initially situated above are marked by points, and those of the fluid initially placed below by circles. We shall note that only some of the markers used in the calculations are printed. In the initial state (at $t = 0$) the system has a horizontal interface passing through the cavity middle. At $G = 6000$ and $\rho = 1.005$ the system at subsequent time moments keeps a two-layer structure. A slight change in a densities ratio completely changes the process: the lower fluid begins to rise gradually in the centre (Fig. 6.4 b–d), then a convective torch appears (Fig. 6.4 e) and finally the fluids mix (Fig. 6.4 f).

Considering the results discussed it is possible to make the following conclusion. There is a region on the plain (G, ρ) (to the right of the dashed curve in Fig. 6.2) in which a two fluids system keeps a two-layer structure both in the reference state and at the convection generation.

2.   Convection in a two-layer system with internal heat sources. Let us discuss now the results of [153] also devoted to the calculation of finite — amplitude convection regimes in a two-layer system with a deformable interface. The fluids fill up a rectangular cavity. In contrast to the case considered in the previous subsection the convection source is heat release in one of the fluids (possessing larger density). Two situations are under consideration: 1. a heavy fluid layer is situated over the light layer ($\rho > 1$); 2. a light fluid layer is situated over the layer of heated heavy fluid ($\rho < 1$). In the first case the upper boundary is assumed to be free ($\partial u/\partial z = v = 0$) and the lower one is assumed to be solid ($u = v = 0$), and in the second case both boundaries are solid. In both cases the horizontal boundaries are kept at a constant temperature ($T = 0$), and on vertical boundaries symmetry conditions are put ($\partial T/\partial x = u = \partial v/\partial x = 0$). Surface tension is taken into account as any volume force, acting only on the interface. The computing technique slightly differs from that described in Section 6.1; particularly, in calculation of the flow at the intermediate stages the stream function is used. At $\rho > 1$ the cause of motion development is Rayleigh–Taylor instability. This instability leads to the interface deformation and separation of the "bubbles". Besides, close to the upper boundary a light fluid layer is formed in which convection arises (see Fig. 6.5 a, b).

The influence of fluids non-isothermicity on the instability development is not large. This fact is proved by the dependences [153] of the boundary disturbance wave length and maximum bubble diameter on the surface tension value which are in good agreement with theoretical and experimental data for Rayleigh–Taylor instability.

At $\rho < 1$ the source of motion is convection caused by nonsteady stratification of the upper fluid. For this case the calculations of steady convective motions in a broad range of the Grashof numbers for differ-

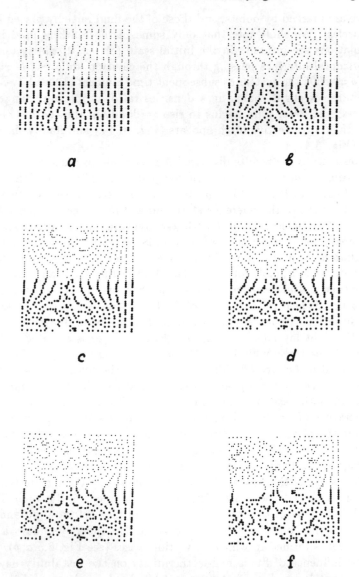

a

b

c

d

e

f

Figure 6.3. Markers position for convection with a two-layer structure.

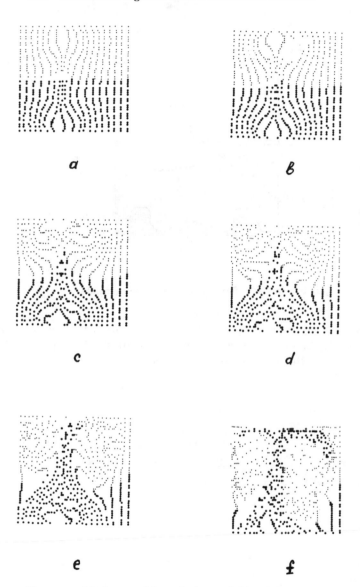

Figure 6.4. Markers position relative to fluids mixing.

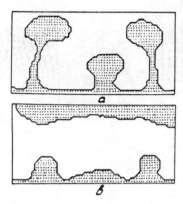

Figure 6.5. Interface form in different time moments.

ent ratios of layers thiknesses and media heat conductivities are carried out. The dependences of heat fluxes ratios through the upper and lower boundaries on the system parameters are investigated.

CHAPTER 7

Effects of Surfactants on the Convective Stability

Absorption of surface-active agents is an important physical effect accompanying convective phenomena in systems with an interface. The surfactant concentrations being small, the convective motion leads to a heterogeneous surfactant distribution on the interface. Owing to the dependence of the surface tension on the surfactant concentration additional tangential stresses arise which essentially influence the convective flow.

It is necessary to note that the interface instability on which the surfactant is adsorbed may be stipulated by the mechanisms which have no relation to heat convection: concentration — capillary effect under mass transfer conditions through the interface [38, 89, 75, 161, 21, 72–74] or in the presence of a heterogeneous chemical reaction [33], dipole interaction of molecules [182], and so on. However, consideration of these phenomena is beyond the limits of our discussion. The present-day state of theory of systems with the interface stability under the conditions of concentration-capillary convection in the presence of chemical and electrical interactions is reflected in collection [78]. Let us also mention experimental investigations of the surfactant effect on mass transfer processes under "surface turbulence" conditions (see article [45] and references there) and also the attempt of the mass transfer turbulent theory construction in the presence of surfactant [84]. As elsewhere, this chapter will be restricted to studying instability types, the development of which is essentially influenced by temperature disturbances. First (Sections 7.1–7.3) the influence of insoluble surfactant on different mechanisms of convective instability (that is, thermocapillary, thermogravitational and mechanism of convection arising when heating from above) is investigated. Further (Sections 7.4 and 7.5) the surfactant solubility is considered.

7.1. Statement of the problem

As far as only linear problems will be considered in this chapter, we shall limit ourselves to a two-dimensional statement. Let surfactant with surface concentration $\Gamma(x)$ concentrate on the interface of finite-thickness immiscible fluids layers. We assume the surfactant concentration to be not large, so that its molecules form a "surface gas". The surfactant transfer along the interface is described by equation [93]:

$$\frac{\partial \Gamma}{\partial t} + \frac{\partial}{\partial x}(v_x \Gamma) = D_0 \frac{\partial^2 \Gamma}{\partial x^2} + j, \tag{7.1}$$

where v_x is the fluid velocity on the interface, D_0 is the coefficient of the surfactant surface diffusion, j is mass flow of dissolved surfactant from the volume to the interface. In Sections 7.1–7.3 $j = 0$ (insoluble surfactant) is assumed. At the reference state the concentration of surfactant on the interface is constant: $\Gamma = \Gamma_0$.

The surfactant influence is shown first in the dependence of the surface tension on concentration which further is approximated by a linear function:

$$\sigma = \sigma_0 - \alpha T - \alpha_s \Gamma, \tag{7.2}$$

(α_s is constant).

Besides, the interface surface unit possesses now the mass Γ, so the stresses balance condition on the interface will be written in the form (the same notation as in Section 1.2)

$$\frac{d}{dt}(\Gamma v_{1,i}) = -(p_1 - \rho_1 gh)n_i + (p_2 - \rho_2 gh)n_i - $$
$$\frac{\sigma}{R}n_i + (\sigma_{1,ik} - \sigma_{2,ik})n_k - \alpha D_i T_1 - \alpha_s D_i \Gamma, \tag{7.3}$$

where $D_i = \partial/\partial x_i - n_i n_k \partial/\partial x_k$ is the surface gradient, $\frac{d}{dt}$ is the full derivative with respect to time. In cases when the interface deformation is not considered (see Section 2.1), the proection of equality (7.3) on the normal to the surface is substituted by the condition $h = 0$, and the tangential component (in a two-dimensional statement) takes the form:

$$\frac{d}{dt}(\Gamma v_{1,x}) = \sigma_{1,xz} - \sigma_{2,xz} - \alpha \frac{\partial T_1}{\partial x} - \alpha_s \frac{\partial \Gamma}{\partial x}. \tag{7.4}$$

Some authors [78, 123] add terms connected with the surface viscosity (see Section 1.2); however, these effects are not considered here. Let us make both equations and boundary conditions dimensionless in the same manner as in Section 1.2; the equilibrium concentration Γ_0 is chosen as the surfactant concentration unit. The dimensionless temperature gradient dT_0/dz at the steady state is equal to $A_1 = -s/(1+\kappa a)$ in the upper fluid and to $A_2 = -s\kappa/(1+\kappa a)$ — in the lower one, where $s = -1$ for heating from above and $s = 1$ for heating from below.

Let us pass on to the stability investigation. Let us impose the stream function disturbances ψ_m, temperature T_m and surfactant concentration Γ on the reference state:

$$(\psi_1, T_1, \psi_2, T_2, \Gamma) = (\tilde{\psi}_1(z), \tilde{T}_1(z), \tilde{\psi}_2(z), \tilde{T}_2(z), \tilde{\Gamma})\exp(ikx - \lambda t).$$

Here k is the wave number, λ is a complex decrement. Further on the sign "tilde" will be omitted.

Linearized convection equations for disturbances have the form (see Section 2.2):

$$c_m D^2 \psi_m - ikGb_m T_m = -\lambda D\psi_m,$$
$$\frac{d_m}{P} DT_m + ikA_m \psi_m = -\lambda T_m \quad (m = 1, 2). \tag{7.5}$$

here $D = d^2/dz^2 - k^2$, $b_1 = c_1 = d_1 = 1$, $b_2 = 1/\beta$, $c_2 = 1/\nu$, $d_2 = 1/\chi$.

Conditions on solid boundaries are,

$$z = 1: \quad \psi_1 = \psi_1' = T_1 = 0,$$
$$z = -a: \psi_2 = \psi_2' = T_2 = 0 \tag{7.6}$$

and

$$z = 0: \quad \psi_1 = \psi_2 = 0, \psi_1' = \psi_2', T_1 = T_2, \kappa T_1' = T_2', \tag{7.7}$$
$$\eta\psi_1'' - ik(MrT_1 + B\Gamma) - \psi_2'' = -\mathcal{K}ik\lambda\psi_1', \tag{7.8}$$

on the interface;

$$M = \frac{\alpha\theta a_1}{\eta_2\nu_1}, \quad B = \frac{\alpha_s\Gamma_0 a_1}{\eta_2\nu_1}, \quad \mathcal{K} = \frac{\Gamma_0}{\rho_1 a_1}\frac{\eta_1}{\eta_2}.$$

It is easy to show that parameter $\mathcal{K}$ is proportional to the ratio of the surfactant particle mass concentrated on the interface to the mass of the first fluid. Further on this parameter is assumed to be small and the expression on the right-hand side of (7.8) is omitted.

Equation (7.1) after transition to dimensionless variables and linearization takes the form

$$(\lambda - D_s k^2)\Gamma = ik\psi_1'(0), \tag{7.9}$$

where $D_s = D_0/\nu_1$.

Eliminating Γ from (7.8) and (7.9), we get boundary condition (at $\mathcal{K} = 0$):

$$z = 0: \quad \eta\psi_1'' - ik\left(MrT_1 + \frac{ikB}{\lambda - D_s k^2}\psi_1'\right) = \psi_2''. \tag{7.10}$$

The stability boundary is defined by equality $\lambda_r = 0$.

7.2.   Thermocapillary convection in the presence of surfactant

The boundary value problem (7.5)–(7.7) and (7.10) describes the convection generation in the system with surfactant at the interface, the Rayleigh and the thermocapillary mechanisms of instability acting simultaneously. Depending on the relationship between parameters Mr and G one or the other mechanism can dominate. In this paragraph we consider the case when thermocapillary effect is the only cause of instability —buoyancy is absent, $G = 0$ [111, 114, 112].

Let us first discuss the results from the theory of thermocapillary instability of one-layer fluid [13, 123]. The surfactant effect is displayed first of all in convection threshold rise [13]. This effect can be explained in the following manner [48]. Let the fluid element influenced by an accidental disturbance come to the interface. At this surface point a region with respectively small surfactant concentration and, hence, with larger surface tension is formed. Therefore, the tangential forces, directed opposite to thermocapillary forces and the fluid motion arise. Thus, the presence of the surfactant adsorbed film must produce a stabilizing effect on the monotonous mode of instability.

The presence of restoring force, as known, may lead not only to stabilization of monotonous instability, but also to the appearance of oscillatory instability [48]. In [123] it has been established for the case of one-layer fluid that oscillatory instability actually arises at sufficiently small relations of the surfactant diffusion coefficient and the coefficient of kinematic viscosity.

Let us return now to the consideration of two-layer systems. It was mentioned in Section 2.5 that, depending on the fluids parameters, the mode of heating and the wave number, the thermocapillary convection may be stipulated both by monotonous and oscillatory disturbances.

Let us first discuss the surfactant effect on the monotonous mode of instability. For this mode ($\lambda_r = \lambda_i = 0$) the boundary value problem (7.5)–(7.7) and (7.10) admits exact solution similar to the solution obtained in Section 2.5 for $B = 0$. The expression for the threshold Mr number has the form

$$Mr = Mr_m = 8sk^2 \frac{1 + \kappa a}{P\kappa} \frac{(\kappa D_1 + D_2)(\eta B_1 + B_2 + B/2kD_s)}{\chi E_2 - E_2}. \quad (7.11)$$

Here the same notation as in (2.57) is used. Let us note that the position of neutral curve breaks defined by the function $\chi E_2(k) - E_1(k)$ noughts is not changed at surfactant appearance. Let us present asymptotics of expression (7.11) in the longwave limit ($k \to 0$) :

$$Mr = -\frac{80s(1 + \kappa a)^2(1 + \eta a + aB/4Ds)}{Pa^2\kappa(1 - \chi a^2)} k^{-2}. \quad (7.12)$$

From (7.11) one can see that the presence of surfactant always leads to the displacement of the monotonous neutral curve towards larger Mr, defined by combination B/D_s. As far as for real surfactants value D_s is usually small, even for moderate values B this displacement is considerable.

As mentioned, the stabilizing effect of monotonous instability in surfactant presence is stipulated by the fact that for monotonous disturbances the tangential forces, caused by temperature and surfactant distribution heterogeneity are of opposite directions [48]. However, it does not exclude an opportunity of oscillatory instability at which the temperature and the surfactant-related disturbances oscillate with a phase shift.

The analysis shows that inclusion of B can really lead to the appearance of an oscillatory neutral curve: $\lambda_r = 0$, $\lambda_i = i\omega$, where ω is the oscillatory frequency. Longwave asymptotics for this curve can be obtained with the help of expansion in parameter k. Let us present the boundary value problem solution at $k \to 0$ as a series

$$\psi_m = \sum_{n=0}^{\infty} k^n \psi_m^{(n)}, \quad T_m = \sum_{n=1}^{\infty} k^n T_m^{(n)},$$

$$\omega = \sum_{n=1}^{\infty} k^n \omega^{(n)}, \quad Mr = \sum_{n=-2}^{\infty} k^n Mr^{(n)}. \quad (7.13)$$

Let as substitute expansions (7.13) in (7.5)–(7.7) and (7.10), and make equal the terms, having the same order in k. In zero order we get:

$$\psi_1^{(0)IV} = 0, \quad i\psi_1^{(0)} \frac{s}{1+\kappa a} = \frac{1}{P} T_1^{(1)''}, \quad 0 < z < 1,$$

$$\psi_2^{(0)IV} = 0, \quad i\psi_2^{(0)} \frac{s\kappa}{1+\kappa a} = \frac{1}{\chi P} T_2^{(1)''}, \quad -a < z < 0,$$

(7.14)

$$z = 1: \quad \psi_1^{(0)} = \psi_1^{(0)'} = T_1^{(1)} = 0,$$

$$z = -a: \quad \psi_2^{(0)} = \psi_2^{(0)'} = T_2^{(1)} = 0,$$

$$z = 0: \quad \psi_1^{(0)} = \psi_2^{(0)}, \quad \psi_1^{(0)'} = \psi_2^{(0)'}, \quad T_1^{(1)} = T_2^{(1)},$$

$$\kappa T_1^{(1)'} = T_2^{(1)'}, \quad \eta \psi_1^{(0)''} - iMr^{(-2)}T_1^{(1)} = \psi_2^{(0)''}.$$

The solution has the form:

$$\psi_1^{(0)} = a^2(z^3 - 2z^2 + z), \quad \psi_2^{(0)} = z^3 + 2az^2 + a^2 z,$$

$$T_1^{(1)} = \frac{isPa^2}{1+\kappa a} \left[\frac{z^5}{20} - \frac{z^4}{6} + \frac{z^3}{6} - \frac{1+\kappa\chi a^3}{20(1+\kappa a)} z - \frac{\kappa a(1-\chi a^2)}{20(1+\kappa a)} \right],$$

$$T_2^{(1)} = \frac{is\kappa\chi P}{1+\kappa a} \left[\frac{z^5}{20} + \frac{az^4}{6} + \frac{a^2 z^3}{6} - \frac{a^2(1+\kappa\chi a^3)}{20\chi(1+\kappa a)} z - \frac{a^3(1-\chi a^2)}{20\chi(1+\kappa a)} \right].$$

The solvability condition of system (7.14) gives the longwave asymptotics of neutral curve for oscillatory instability

$$Mr = Mr_Q \simeq Mr^{(-2)}k^{-2} = -\frac{s}{1-\chi a^2} \frac{80(1+\kappa a)^2(1+\eta a)}{Pa^2\kappa} k^{-2}.$$

(7.15)

In the first order in k we get the boundary value problem:

$$i\omega^{(1)}\psi_1^{(0)''} = -\psi_1^{(1)IV}, \quad -i\omega^{(1)}T_1^{(1)} + i\psi_1^{(1)} \frac{s}{1+\kappa a} = \frac{1}{P} T_1^{(2)''}, \quad 0 < z < 1,$$

$$i\omega^{(1)}\psi_2^{(0)''} = -\frac{1}{\nu}\psi_2^{(1)IV}, \quad i\omega^{(1)}T_2^{(1)} + i\psi_2^{(1)} \frac{s\kappa}{1+\kappa a} = \frac{1}{\chi P} T_2^{(2)''}, \quad -a < z < 0,$$

$$z = 1: \quad \psi_1^{(1)} = \psi_1^{(1)'} = T_1^{(2)} = 0,$$

$$z = -a: \quad \psi_2^{(1)} = \psi_2^{(1)'} = T_2^{(2)} = 0,$$

$$z = 0: \quad \psi_1^{(1)} = \psi_2^{(1)} = 0, \quad \psi_1^{(1)'} = \psi_2^{(1)'},$$

$$T_1^{(2)} = T_2^{(2)}, \quad \kappa T_1^{(2)'} = T_2^{(2)'},$$

$$i(\eta\psi_1^{(1)''} - \psi_2^{(1)''} - iMr^{(-2)}T_1^{(2)} - iMr^{(-1)}T_1^{(1)})\omega^{(1)} + B\psi_1^{(0)} = 0,$$

the solvability condition of which defines the asymptotics of frequency:

$$\omega \simeq \omega^{(1)}k, \omega^{(1)} = \pm B^{1/2}a\left\{\frac{2}{15}(\eta a^2 + \nu a^3) + \frac{1}{315}\frac{1 + a\eta}{1 - \chi a^2}\cdot\right.$$

$$\left[10P\frac{11(1 - \kappa\chi^2 a^5) + 53a(\kappa - \chi^2 a^3) + 42\chi a^2(1 - \kappa a)}{1 + \kappa a} + \right. \tag{7.16}$$

$$\left.\left.19(1 - \nu\chi a^4)\right]\right\}^{-1/2}$$

Oscillatory instability arises in the region $k \to 0$ when the expression (7.16) is real-valued.

Let us note that in the longwave limit the form of oscillatory neutral curve does not depend on parameter B. Comparing (7.12) and (7.15) one can see that oscillatory neutral curve at $B \neq 0$, $k \to 0$ always lies lower than the monotonous one. Let us pay attention to the difference between the longwave asymptotics of oscillations frequency caused by surfactant ($\omega = \omega^{(1)}k$) and the asymptotics for thermocapillary oscillations at $B = 0$ ($\omega = $ const at $k \to 0$; see Section 2.5).

To calculate oscillatory neutral curves for finite k, the numerical Runge–Kutta method was used. Let us consider the water-silicone oil DC N 200 system (see Table 1), on the interface of which surfactant is deposited. All calculations of this paragraph were carried out for dimensionless parameter value $D_s = 10^{-3}$. Let $a = 1$. The surfactant being absent, only monotonous thermocapillary instability when heating from above takes place (see Section 2.5). At $B \neq 0$ the monotonous neutral curve moves towards larger Mr, according to (7.11). An oscillatory neutral curve arises in the longwave region $(0 < k < k_*(B))$. In the limit $k \to 0$ the neutral curve form and the oscillations frequency are described by (7.15), (7.16). At $k \to k_*(B)$ the oscillations frequency goes to zero and the oscillatory neutral curve is terminated at the monotonous

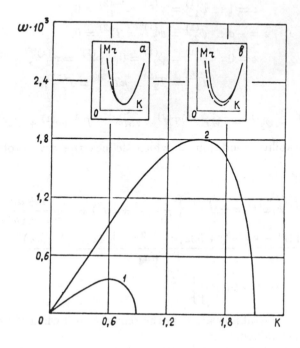

Figure 7.1. Dependence of the convective oscillations frequency and threshold Mr number on the wave number.

one. In Fig. 7.1 graphs of dependence $\omega(k)$ for $B = 10^{-5}$ (line 1) and $B = 4 \cdot 10^{-5}$ (line 2) are shown. For the same values of B in the inserts a and b, respectively, the monotonous (solid lines) and the oscillatory (dashed lines) neutral curves are presented. Functions of $Mr(k)$ are shown qualitatively because at these values of B the lines merge in the graph scale. Let us underline that the part of monotonous neutral curve in the region $k < k_*$ situated higher than oscillatory neutral curve can not be considered as the boundary of monotonous instability.

In Fig. 7.2 neutral curves for different values of B are presented. Monotonous neutral curves for $B \neq 0$ lie at considerably higher Mr values and are not shown in the graph.

Fig. 7.3 illustrates the typical form of decrement dependence λ on Mr in the region $k < k_*$; dashed line corresponds to the oscillatory, and solid line to the monotonous disturbances. With the increase in Mr number,

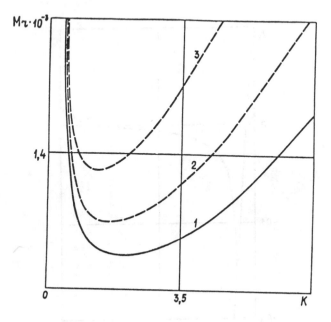

Figure 7.2. The neutral curves for the values of parameter $B=0$ (line 1); 3 (2); 8 (3); $a=1$.

oscillatory instability arises at $Mr = Mr_o = 380$; at $Mr > Mr_d = 507$ the oscillations frequency vanishes, and the system possesses two types of monotonously increasing disturbances. At $Mr > Mr_m = 2200$ one of the monotonous disturbances types becomes decreasing. That is why a rapid Mr_m increase with the increase in B cannot be interpreted as stabilization of monotonous instability.

The neutral curves for thermocapillary instability are of more complicated character if surfactant is absent ($B = 0$) at $a = 2$. In this case the graph of the function $sMr(k)$ for monotonous mode (line 1, Fig. 7.4) has a break at $k = k_1 \simeq 1.45$: as $k < k_1$ the monotonous instability is realized when heating from below ($sMr > 0$) and at $k > k_1$ — when heating from above ($sMr < 0$). In the range $k < k_* \simeq 1.72$ when heating from above the instability is of the oscillatory character (line 2, Fig. 7.4).

When heating from below ($sMr > 0$) the presence of surfactant leads to the neutral curves' splitting, as in the case $a = 1$ discussed above; the longwave frequency asymptotics has the form $\omega \sim k$. For heating from above ($sMr < 0$) with the increase in B the final point of the

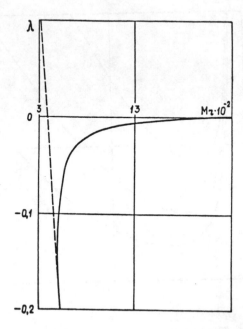

Figure 7.3. Dependence of disturbances decrement on Mr; $B=0.05$; $k=2.4$.

oscillatory neutral curve on the monotonous one $k = k_*$ displaces to the larger k direction. The longwave frequency asymptotics keeps the form $\omega = $ const, being identical to that at $B = 0$.

Fig. 7.4 presents oscillatory neutral curves in the presence of surfactant. The monotonous curves lie at considerably larger values of Mr and are not presented in the graph. The dependence of frequency on the wave number for the same cases is shown in Fig. 7.5. Graphs of $\omega(k)$ close to $k = k_*(B)$ at small B for heating from above are shown in Fig. 7.6.

For other systems the neutral curves have the form similar to the presented in Fig. 7.2 (if convection arises only when heating from one side) or in Fig. 7.4 (if convection is possible for both variants of heating).

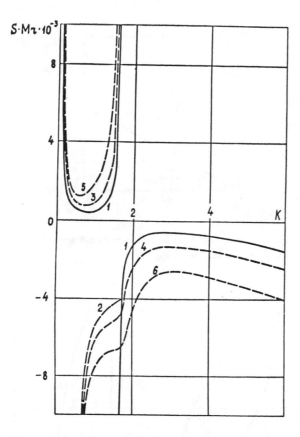

Figure 7.4. The neutral curves for the values of parameter $B=0$ (lines 1,2); 3 (3,4); 8 (5,6); $a=2$.

7.3. Thermogravitational convection

Let us consider the case, when thermocapillary effects are not essential $(Mr \ll G)$. In this case in the boundary condition (7.10) it is possible to set $Mr = 0$.

If the system is heated from below, Rayleigh–Taylor instability arises in it. As shown in Section 2.2, in the absence of surfactant $(B = 0)$ convective instability of the steady state may arise both in monotonous and in oscillatory manner, the monotonous instability being more typical; up to now the oscillatory instability has been found out only for one real fluid system (transformer oil-formic acid) in a limited interval

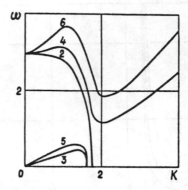

Figure 7.5. Dependences of oscillations frequency on the wave number. Lines numeration relates to Fig. 7.4.

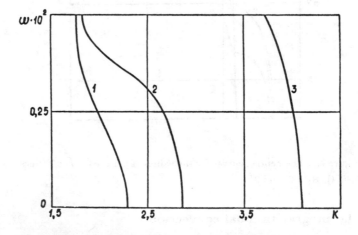

Figure 7.6. Dependences of oscillations frequency on the wave number in the neighbourhood of the oscillatory neutral curve final point for the values of parameter $B=4\cdot10^{-5}$ (line 1); $6\cdot10^{-5}$ (2); 10^{-4} (3).

of layers' thickness ratio a, the neutral curve minimum being realized for monotonous disturbances.

1.   Surfactant effect on monotonous mode of instability (heating from below). Let us first discuss the surfactant influence on the thermogravitational convection onset in the case when in the absence of surfactant the instability is monotonous. As an example the air-water system is chosen (see Table 1; $a = 1$, $D_s = 10^{-3}$) [58]. Calculations show that for this system convective instability of the reference state in the presence of surfactant may arise both in monotonous and in oscillatory manner. Let us discuss the monotonous instability case ($\lambda = 0$). In Fig. 7.7 the monotonous neutral curves for different values of B are shown in solid lines. Dependence of the minimized (according to the wave number) critical Grashof number G_* on B is shown in Fig. 7.8. From the boundary condition (7.10), it is clear the characteristic parameter of the problem is the ratio B/D_s, already at small values of $B \sim 10^{-1}$ the neutral curve becomes close to the limiting one ($B \to \infty$). This limit corresponds to the switch of the boundary condition (7.10) to the condition $\psi_1'(0) = 0$, physically corresponding to a solid division between the media.

The surfactant being included, oscillatory instability appears. At small B oscillations appear in the longwave region $0 < k < k_*(B)$. As B increases, the final point of the oscillatory neutral curve moves into the shortwave region, and any value B_1 exceeding, the oscillatory disturbances become the most dangerous. As B increases further, $k_*(B)$ begins to decrease and as a result (at $B > B_2$) the monotonous disturbances again become the most dangerous. We shall note that at $B = \infty$ (solid interface) the boundary value problem becomes self-adjoint independent on the fluids parameters, that is why the oscillatory instability is impossible. In Fig. 7.7 oscillatory neutral curves are shown by dashed lines. The dependence of the minimized critical Grashof number $G_*(B)$ for oscillatory instability is shown by dashed line in Fig. 7.8. Fig. 7.9 presents dependences of the frequency ω on the wave number k.

Thus, the presence of surfactant on the interface causes oscillatory instability of a specific type which under definite conditions turns out to be the most dangerous. It is possible to show analytically (with the help of expansions in parameter k), that longwave asymptotics of oscillatory neutral curve $G_o(k, B)$ at $B \neq 0$ does not depend on B and absolutely coincides with asymptotics of monotonous neutral curve $G_m(k, 0)$ as $B = 0$:

$$\lim_{k \to 0} G_o(k, B)k^2 = \lim_{k \to 0} G_m(k, 0)k^2 = \text{const.}$$

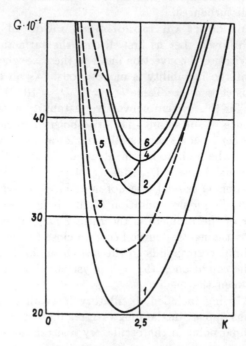

Figure 7.7. Monotonous and oscillatory neutral curves for the values $B=0$ (line 1); 0.02 (2,3); 0.05 (4,5); 0.09 (6,7).

Longwave asymptotics of the frequency at $k \to 0$ has the form

$$\omega = k\sqrt{BC} + O(k)$$

where C is a function of the system parameters. It is necessary to under-line that, like in case of thermocapillary instability (see Section 7.2), the part of monotonous neutral curve in the region $k < k_*(B)$ situated higher than the oscillatory neutral curve (see Fig. 7.7) is not the boundary of monotonous instability. In Fig. 7.10 the typical form of the decrement λ dependence on G in the region $k < k_*$ is shown; the dashed line corre-sponds to oscillatory and the solid one — to monotonous disturbances. As G number increases, oscillatory instability arises at $G = G_0 = 445$; at $G > G_d = 470$ the oscillations frequency vanishes and the system has two types of monotonously increasing disturbances, one of which at $G > G_m = 471$ starts to decrease (compare with Fig. 7.3).

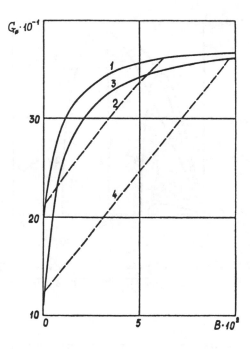

Figure 7.8. Dependences of the minimized critical Grashof number on parameter B for $Mr=0$ (lines 1,2), 10 (3,4).

Let us discuss now the influence of the thermocapillary effect. In the absence of surfactant with the increase in Mr number the critical Grashof number G_* decreases, but instability remains to be monotonous. In the presence of surfactant, like in the case of pure gravitational convection ($Mr = 0$), two neutral curves exist: a monotonous and an oscillatory one. In Fig. 7.8 the critical Grashof number $G_*(B)$ for monotonous (line 3) and oscillatory (line 4) instabilities at $Mr = 10$ are shown. Let us note that the interval of B values in which the oscillatory disturbances are the most dangerous increases with the increase in Mr. Fig. 7.11 presents the dependences of (minimized with respect to k) the Grashof number G_* on Mr number at fixed values of B for the monotonous (solid lines) and the oscillatory (dashed lines) neutral curves.

2. The surfactant influence on the oscillatory mode of instability (heating from below). Let us describe now the surfactant

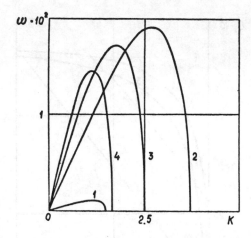

Figure 7.9. Dependences of the frequency on the wave number for the values of parameter $B=0.0005$ (line 1); 0.02 (2); 0.05 (3); 0.09 (4).

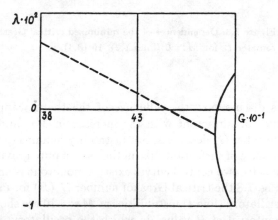

Figure 7.10. Dependence of the disturbances decrement on the Grashof number; $B=0.09$; $k=1.5$.

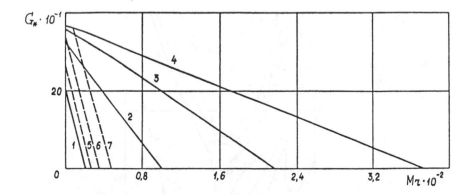

Figure 7.11 Dependences of the minimized Grashof number on Mr number for $B{=}0$ (line 1); 0.02 (2,5); 0.05 (3,6); 0.09 (4,7).

influence on the convection generation in the transformer oil-formic acid system with parameters presented in Table 1; $a = 0.667$.

Convective marginal stability for this system in the absence of surfactant ($B = 0$) is considered in Section 2.2. Outside of the interval of wave numbers $2.3 < k < 3.6$ the instability has a monotonous character (see Fig. 7.12), and it is possible to differentiate instability boundaries on which convection arises mainly in the oil layer (lines 1 and 4) and in the acid layer (lines 2 and 3). A "circuit" of monotonous neutral curves takes place on the boundaries of the indicated interval causing an oscillatory neutral curve (line 5). At small values of B close to either boundary of monotonous instability (lines 1 and 2), corresponding to the convection onset in the oil layer and in the acid layer in the longwave region the zones of oscillatory instability stipulated by the surfactant arise; the mechanism of their appearance does not differ from that described above. As B increases, one of these parts unites with the oscillatory neutral curve which existed in the absence of surfactant, forming a united oscillatory neutral curve, crossing with the monotonous one. The neutral curves picture takes the form shown in Fig. 7.12, 7.13 (lines 6–11, $B = 0.05$).

As B further increases, a pair of monotonous neutral curves appears. Let us describe the picture of neutral curves at $B = 0.06$. Lower branches of both monotonous neutral curves unite, forming a united monotonous neutral curve (line 12); the same phenomenon takes place for the upper branches of neutral curves (line 13). The oscillatory neutral curve (line

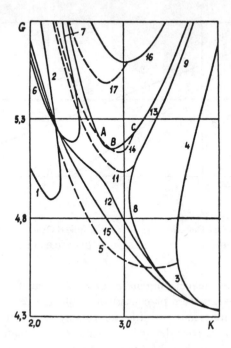

Figure 7.12. Picture of neutral curves for a transformer oil-formic acid
system; a=0.667.

14) crosses the boundary of monotonous instability 13 at points A and
B, and terminates on this boundary at point C at which the oscillations
frequency vanishes. The region limited from above by the oscillatory
(line 14) and from below — by the monotonous (line 13) (between the
points A and B) is the "stability islet". With the increase in B, the
picture of neutral curves simplifies (lines 15–17, $B = 0.1$).

**3. The surfactant effect on convective stability when heating
from above.** In Section 2.3 it is stated that when heating from above
in the absence of surfactant for some two-layer systems the monotonous
instability stipulated by specific non-Rayleigh mechanism [49, 55] is possible. Physical mechanism of instability has been explained on the ex-
ample of the system with $\chi_2 \rightarrow \infty$, $\beta_1 \rightarrow 0$ for which the instability
boundary was calculated analytically.

Let us consider the stability of this model system in the presence of
surfactant on the interface [115]. The Grashof number G determined by

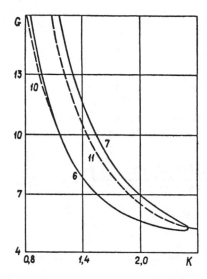

Figure 7.13. The neutral curves in a longwave region.

the parameters of the upper fluid at $\beta_1 = 0$ vanishes. The model system being studied, we shall introduce, as done in Section 2.3, the Grashof number defined by the parameters of the lower fluid

$$G' = \frac{g\beta_2 \theta a_2^3}{\nu_2^2}. \tag{7.17}$$

In the limit under consideration equations (7.5) for $\lambda = 0$ (stability boundary) after replacement of variables $T_m = P\theta_m$ ($m = 1, 2$) are reduced to the form

$$D^2\psi_1 = -i\omega D\psi_1, \quad D\theta_1 + ikA_1\psi_1 = -i\omega P\theta_1,$$
$$D^2\psi_2 - ikR'\theta_2 = -i\omega\nu D\psi_2, \quad D\theta_2 = 0, \tag{7.18}$$

where R' is the effective Rayleigh number $R' = G'P/\nu a^3$, $A_1 = 1/(1 + \kappa a)$. The influence of thermocapillary effect considered to be of any small value ($Mr = 0$), so we shall write the boundary conditions in the form:

$$z = 1: \quad \psi_1 = \psi_1' = \theta_1 = 0,$$
$$z = -a: \quad \psi_2 = \psi_2' = \theta_2 = 0,$$
$$z = 0: \quad \psi_1 = \psi_2 = 0, \psi_1' = \psi_2', \theta_1 = \theta_2, \kappa\theta_1' = \theta_2',$$
$$\eta\psi_1'' + \frac{k^2 B}{\lambda - D_s k^2}\psi_1' = \psi_2''.$$

Let us consider first the monotonous instability case. The above-stated boundary value problem determines the neutral curve $R'(k)$:

$$R'(k) = \frac{32(1 + \kappa a)k^4(\kappa S_2 C_1 + C_2 S_1)}{\kappa h_1 h_2}\left(\eta f_1 g_2 + f_2 g_1 + \frac{B}{2kD_s}g_1 g_2\right),$$
$$(7.19)$$

where

$$S_1 = \text{sh}k, C_1 = \text{ch}k, S_2 = \text{sh}ka, C_2 = \text{ch}ka,$$
$$f_1 = S_1 C_1 - k, f_2 = S_2 C_2 - ka, g_1 = S_1^2 - k^2,$$
$$g_2 = S_2^2 - k^2 a^2, h_1 = S_1^3 - k^3 C_1, h_2 = S_2^3 - k^3 a^3 C_2.$$

One can see that the presence of surfactant leads to the instability threshold increase; the characteristic problem parameter is ratio B/D_s. In contrast with the above described Rayleigh convection for which at $B \to \infty$ the threshold Rayleigh number goes to any finite value corresponding to the solid interface, when heating from above in the limit $B \to \infty$, $R'(k)$ grows indefinitely.

The problem of the oscillatory instability existence in a longwave region can be studied analytically. In Subsection 2 of the given paragraph for the case of Rayleigh convection the longwave asymptotics of the oscillation frequency were obtained

$$\omega = ck, \quad c^2 = Bf, \quad (7.20)$$

where f is a function of the system parameters. Expression (7.20) is also applicable in the case when heating from above. For the model system under consideration we get:

$$f = 315a\Big[(19 + 61\eta a) + \nu a^2(61 + 19\eta a)+$$

$$10(1 + \eta a)\left(25 + 21\frac{1 + 3\kappa a}{1 + \kappa a}\right)\Big]^{-1}. \quad (7.21)$$

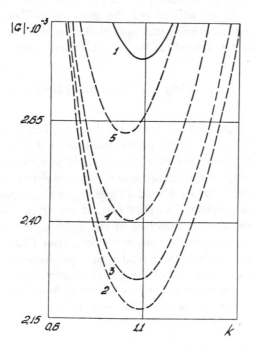

Fig. 7.14. The neutral curves for a water-mercury system when heating from above: $B=0.001$ (lines 1,2); 0.5 (3); 1.5 (4); 1 (5); $a=1$; $D_s=10^{-3}$.

One can see that variable f is always positive, so the oscillatory instability of the steady state is realized for any values of parameters η, ν, κ, a and P. The analysis of the longwave asymptotics of the monotonous and the oscillatory neutral curves shows that oscillatory instability in the longwave region turns out to be the most dangerous.

Fig. 7.14 presents numerical calculations results of neutral curves for air-mercury system at $20°C$ (parameters κ, χ, η are presented in Table 2; $P = 7.17$; $\nu = 8.81$; $\beta = 1.14$). The presence of surfactant leads to the considerable stabilization of monotonous instability; the effect of stabilization is defined by combination B/D_s. An oscillatory neutral curve arises in the longwave region. As B increases, the curve stabilizes in a more weak way than the monotonous one. Even at $B = 0.001$ the oscillatory instability (line 2, Fig. 7.14) becomes more dangerous than the monotonous one (line 1). Owing to this, the monotonous neutral curves for $B > 0.001$ are not shown in the graph.

Thus, the results of Sections 7.2, 7.3 prove that the stabilization effects of monotonous instability and the onset of oscillatory instability stipulated by the presence of surfactant at the interface have a universal character and do not depend on the physical mechanism of instability.

7.4.   Effects of solubility on thermocapillary convection

In Section 7.1 the problem of thermocapillary convection generation in a two-layer system was solved under assumption that the surfactant placed on the interface is insoluble. Let us study now the influence of the surfactant solubility on the reference stability in a system with interface [116]. In this section the Archimedes force is not taken into account.

To simplify the procedure, we shall restrict ourselves to the case when the surface-active addition dissolves in the lower media only. The admixture is characterized by the volume concentration $C(x, z)$ and surface concentration $\Gamma(x)$. Thermodynamically, the equilibrium surface concentration Γ_e value is a function of the variables of volume concentration C, temperature T_2 and pressure p_2 in the lower fluid $\Gamma_e = f(C, T_2, p_2)$, the explicit form of which is not used. At small deviation of Γ from Γ_e the admixture flow arises

$$j = -\frac{1}{\tau}[\Gamma - f(c, T_2, p_2)], \qquad (7.22)$$

where τ is relaxation time ($j > 0$ if the flow is directed from the balk to the surface). The change of the surfactant surface concentration with convection, diffusion, adsorption and desorption taken into account is described by equation (7.1). The distribution of volume concentration C is defined by the boundary value problem[1]

$$\frac{\partial C}{\partial t} + v_2 \nabla C = d_c \Delta C, \quad -a_2 < z < 0,$$

$$z = -a_2 : \quad \frac{\partial C}{\partial z} = 0, \quad z = 0 : \quad -D_c \frac{\partial C}{\partial z} = j; \qquad (7.23)$$

where D_c is the coefficient of admixture volume diffusion.

---

[1] Thermodiffusion phenomenon, causing the concentration gradient if a temperature gradient is imposed (the Soret effect) is not considered here. In one-layer approach convection in the presence of a soluble surfactant at a free surface was studied in [30].

Mechanical reference state is characterized by the constant value of vertical temperature gradient equal to $-s\theta\kappa_2/(a_1\kappa_2+a_2\kappa_1)$ in the upper medium and $-s\theta\kappa_1/(a_1\kappa_2+a_2\kappa_1)$ — in the lower medium ($s=-1$ for heating from above, $s=1$ for heating from below), and also by the constant values of surface and volume concentration of surfactant: Γ_0 and C_0.

Passing on to the investigation of stability, let us discuss the problem of surface tension being dependent on the volume and surface admixture concentration. In the thermodynamic equilibrium state, as noted, the surface concentration and the value of volume concentration near the surface are simply connected. That is why there is no difference whether we consider the surface tension of volume concentration function C (at small C this function may be considered linear: $\sigma=\sigma_0-\alpha T-\alpha_c C$, α_c is constant), or surface concentration Γ (formula (7.2)). Coefficients α_c and α_s are connected by

$$\alpha_c = \alpha_s e, \qquad (7.24)$$

where

$$e = \left(\frac{\partial f}{\partial c}\right)_{T_2, p_2}$$

For slightly-nonequilibrium system state out of equilibrium correlation between Γ and C, the concept of the surface tension can be also used. This time, however, the surface tension depends on the two independent variables C and Γ; at small C and Γ

$$\sigma = \sigma_0 - \alpha T - \tilde{\alpha}_s\Gamma - \tilde{\alpha}_c C; \qquad (7.25)$$

the variables α_s, α_C are expressed through $\tilde{\alpha}_s$, $\tilde{\alpha}_C$ by equations

$$\alpha_s = \tilde{\alpha}_s + \tilde{\alpha}_c e^{-1}, \quad \alpha_c = \tilde{\alpha}_c + \tilde{\alpha}_s e.$$

Transition to dimensionless variables and definition of boundary conditions for small equilibrium disturbances will be performed in the same way as in Section 7.1; for volume concentration and flow density units we shall choose, respectively, Γ_0/e, $\Gamma_0\nu_1/ea_1$. Let us impose the disturbances of stream function ψ_m, temperature T_m, volume concentration C, surface concentration Γ and mass flow j on the reference state:

$$(\psi_1, T_1, \psi_2, T_2, C, \Gamma, j) =$$
$$(\tilde{\psi}_1(z), \tilde{T}_1(z), \tilde{\psi}_2(z), \tilde{T}_2(z), \tilde{C}(z), \tilde{\Gamma}, j)\exp(ikx - \lambda t).$$

In linearized equations (7.5) for stream function disturbances and temperature further we assume $G = 0$. Boundary conditions (7.6) and (7.7) are kept, and (7.8) takes the form

$$\eta\psi_1'' - ik(MrT_1 + \tilde{B}\Gamma + \tilde{B}_cC) - \psi_2'' = -ikK\lambda\psi_1', \qquad (7.26)$$

where

$$\tilde{B} = \frac{\tilde{\alpha}_s\Gamma_0 a_1}{\eta_2\nu_1}, \quad \tilde{B}_c = \frac{\tilde{\alpha}_c\Gamma_0 a_1}{\eta_2\nu_1 e} \quad (B = \tilde{B} + \tilde{B}_c).$$

Linearizing equation (7.22) we get:

$$z = 0 : j = -AE(\Gamma - C - k_T T_2 - k_p p_2), \qquad (7.27)$$

where

$$E = \frac{e^2}{\nu_1\tau}, \quad A = \frac{a_1}{e},$$

$$k_T = \left(\frac{\partial f}{\partial T_2}\right)_{c,p_2}\frac{\theta}{\Gamma_0}, \quad k_p = \left(\frac{\partial f}{\partial p_2}\right)_{c,T_2}\frac{\rho_1\nu_1^2}{\Gamma_0 a_1^2}$$

(for the pressure unit the value $\rho_1\nu_1^2/a_1^2$ is chosen). Here E is a dimensionless parameter which does not depend on geometrical system dimensions, and characterise relaxation time, and parameter A is proportional to the layer thickness. Further on the dependence of the surface concentration equilibrium value on temperature and pressure is neglected.

Equations and boundary conditions (7.1) and (7.23) in dimensionless variables after linearization are reduced to the form

$$-\lambda C = Sc^{-1}DC, \qquad (7.28)$$

$$z = -a : \quad C' = 0, z = 0 : \quad C = -Scj,$$

$$z = 0 : \quad -\lambda\Gamma + ik\psi_1' = -D_s k^2\Gamma + Aj, \quad D_s = \frac{D_0}{\nu_1}, \qquad (7.29)$$

where $Sc = \nu_1/D_c$ is the Schmidt number.

The boundary value problem (7.28), (7.29) with (7.27) taken into account admits analytical solution

$$C(z) = \frac{AE Sc\Gamma \operatorname{ch}[q(z+a)]}{q\operatorname{sh}qa + AE Sc\operatorname{ch}qa},$$

$$q = (k^2 - \lambda Sc)^{1/2}.$$

Substituting in (7.27) we find

$$j = -\frac{AE\Gamma}{1 + AE Sc\operatorname{cth}(qa)q^{-1}}.$$

Eliminating Γ from (7.8), (7.29), we get (at $\mathcal{K} = 0$) :

$z = 0 :$

$$\eta\psi_1'' - ikMrT_1 + k^2\left[\tilde{B} + \frac{\tilde{B}_c}{1 + (AE Sc^{-1}q\operatorname{th}qa}\right] \times$$

$$\left(\lambda - D_s k^2 - \frac{A^2 E}{1 + AE Scq^{-1}\operatorname{cth}qa}\right)^{-1}\psi_1' = \psi_2''. \quad (7.30)$$

Problem (7.5)–(7.7), (7.30) describes both the monotonous and the oscillatory instability branches. For the boundary of monotonous instability ($\lambda = 0$) it is possible to get an analytical expression, generalizing similar criteria for insoluble surfactant (see (7.11)):

$$Mr = 8sk^2\frac{1 + \kappa a}{P\kappa}\frac{(\kappa D_1 + D_2)[\eta B_1 + B_2 + \mathcal{B}(k)/2(D_s k + A^2 E\mathcal{F}(k))]}{\chi E_2 - E_1},$$

$$(7.31)$$

where

$$\mathcal{B}(k) = \tilde{B} + \tilde{B}_c\frac{AE Sc\operatorname{cth}ka}{k + AE Sc\operatorname{cth}ka},$$

$$\mathcal{F}(k) = (k + AE Sc\operatorname{cth}ka)^{-1}.$$

In (7.31) the notation of Section 2.5 is used.

Dimensionless parameters A, E, Sc, characterizing the processes connected with surfactant solubility enter formula (7.31) only in combinations

$$AESc = \frac{a_1 e}{D_c \tau}$$

and

$$A^2 E = \frac{a_1^2}{\nu_1 \tau}.$$

Let us discuss the question of the limiting transition to the case of insoluble surfactant. Admixture insolubility may be understood in two ways.

1. The volume concentration of admixture is small, that is why small changes in volume concentration correspond to considerable changes of the surface concentration. This case corresponds to limit $e \to \infty$, i.e. $B = \tilde{B}$, $\tilde{B}_c = 0$, $AESc \to \infty$, $A^2 E = $ const. In this case formula (7.31) becomes (7.11).

2. Thermodynamically, equilibrium volume concentration of admixture is not small, but the stabilization time of equilibrium between volume and surface concentration is large: $\tau \to \infty$. In this case $AESc \to 0$, $A^2 E \to 0$, formula (7.31) goes over into a formula which differs from (7.11) in that B is replaced with $\tilde{B}$.

The same result is obtained in the limit of thin fluid layers: $A \to 0$, $E = $ const. In the opposite limit $A \to \infty$, $E = $ const the critical value of Mr number is the same as for the case of thermocapillary convection in the absence of surfactant.

For disturbances with rather large wave length ($k \text{thk} a \ll AESc$) $\mathcal{B}(k) \simeq \tilde{B} + \tilde{B}_c = B$, $\mathcal{F}(k) = (AESc)^{-1} \text{thk} a$. In this case the solubility effect ($A \neq 0$) is reduced to the effective decrease of parameter B :

$$Mr(A, B, k) = Mr(O, B_e(k), k),$$

$$B_e(k) = \frac{B}{1 + A(k \text{cthk} a)^{-1} H}, \quad B = \tilde{B} + \tilde{B}_c, \quad H = (D_s Sc)^{-1}.$$

It shows that parameter $H = D_c/D_0$ dominates the velocity of monotonous instability threshold decrease, as A increases. At $k \to \infty$ $\mathcal{B}(k) \to \tilde{B}$, $\mathcal{F}(k) \to 0$, i.e. the shortwave asymptotics of neutral curves does not depend on the parameters characterizing the surfactant solubility.

Let us pass on to the consideration of oscillatory instability. In Section 7.1 for the case of insoluble surfactant the longwave asymptotics of oscillations frequency were obtained (see (7.16))

$$\omega = \omega^{(1)}k, \omega^{(1)2} = Bf(P, \eta, \nu, \kappa, z\chi, a).$$

Here f is the known function of system parameters taking the positive value, particularly, for the air-water system at $a = 1$. The longwave expansions similar to those derived in Section 7.1 lead for the soluble surfactant to the equation

$$\omega^{(1)2} = \frac{Bf}{1 + aA}, \quad B = \tilde{B} + \tilde{B}_c. \tag{7.32}$$

It follows that longwave oscillations frequency decreases with the increase in A, but in the longwave region the oscillatory instability is kept at any values of A.

Let us describe the numerical results obtained for the air-water system (see Table 1).

Monotonous neutral curves were constructed on the basis of exact formula (7.31). The calculation of oscillatory neutral curves is carried out numerically by the Runge–Kutta method for $a = 1$, $D_s = 10^{-3}$, $E = 10^2$, $B_c = 0$.

Let us first discuss the case $Sc = 10^2$ ($H = 10$). Fig. 7.15 presents neutral curves for $A = 1$ (curves 1 and 2); $A = 1.2$ (curves 3, 4); $A = 1.4$ (curves 5 and 6). An essential destabilization of the monotonous mode takes place even at $A \sim 1$. The oscillatory neutral curve with the increase in A moves towards larger Mr numbers. At $A \simeq 1.3$ the monotonous instability becomes the most dangerous. Frequency ω as a function of wave number k for the same values of parameter A (curves 1–3) is shown in Fig. 7.16. At the final points of the oscillatory neutral curve on the monotonous one the oscillations frequency vanishes. In Fig. 7.17 neutral curves as $Sc = 10^3$ for the values of parameter $A = 0$ (curve 1); 4.5(2); 13.5(3); 53.5(4,5); 73.5(6,7) are presented. In this case $H = 1$, that is why destabilization of the monotonous mode with the increase in A is considerably slower. We shall note that as A increases, the oscillatory neutral curve displaces first towards the larger and then towards the smaller Mr numbers. Fig. 7.18 presents frequency ω as a function of the wave number k for the same values of parameter A as in Fig. 7.17 (curves 1–5). Thus, the evolution of neutral curves with the increase in parameter A essentially depends on the volume and surface surfactant diffusion coefficients. If the variable H is large, then even at small values

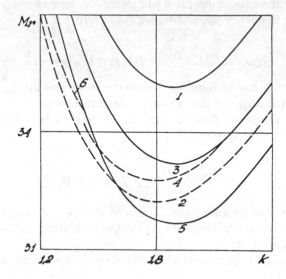

Figure 7.15. The neutral curves for an air-water system (heating from below); $Sc=10^2$.

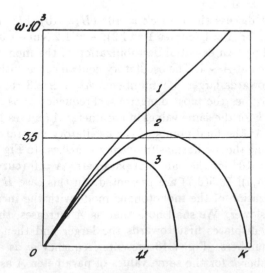

Figure 7.16. Dependences of the oscillations frequency on the wave number $(Sc=10^2)$.

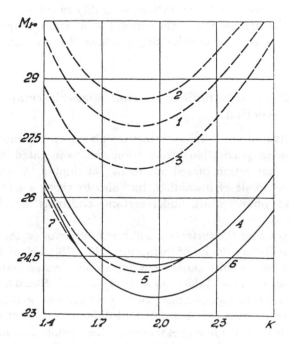

Figure 7.17. The neutral curves ($Sc=10^3$).

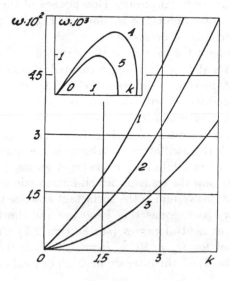

Figure 7.18. Dependences of the oscillations frequency on the wave number ($Sc=10^3$).

of parameter A the oscillatory instability is replaced by the monotonous one. On the contrary, the surfactant solubility may lead to the lowering of the oscillatory instability threshold. But at rather large values of A the monotonous mode of instability becomes the most dangerous.

7.5.   Effects of solubility on thermogravitational convection

The influence of insoluble surfactant placed on the interface on the onset of thermogravitational convection was investigated in Section 7.3. The surfactant action turned out to be not limited by stabilization of monotonous mode of instability, but also to cause a new type of oscillatory instability which under certain conditions may be the most dangerous one.

Let us consider the surfactant solubility influence on the monotonous and the oscillatory modes of instability [117, 118]. Let us first describe the calculations results carried out for an air - water system with parameters presented in Table 1 ($a = 1$; $D_s = 10^{-3}$; $B = 0.02$; $E = 10^2$).

In Fig. 7.19 the monotonous (solid lines) and the oscillatory (dashed lines) neutral curves for $Sc = 10^3$ and different values of parameter A are presented. As parameter A increases, the monotonous neutral curves are seen to lower, narrowing the region of the oscillations existence. The final point of the oscillatory neutral curve on the monotonous one shifts towards the longwave region. Any A_* value exceeding, the monotonous mode becomes the most dangerous. Dependence of the oscillations frequency ω on the wave number k is shown in Fig. 7.20; as A increases, the oscillations frequency decreases.

A similar graph for the case $Sc = 10^2$ is shown in Fig. 7.21 and 7.22. Calculation results at $S = 10^3$ being compared, it follows that decrease in Sc (increase of the volume diffusion coefficient) leads to more effective suppression of oscillations with the increase in A.

The dependence of minimized according to k the Grashof number G on parameter B is presented in Fig. 7.23. The region of parameter B values in which the oscillatory instability is the most dangerous one becomes narrow when solubility parameter A grows.

Let us consider now the transformer oil-formic acid system (see Table 1; $a = 0.667$). For this system in the surfactant absence the neutral curve has the oscillatory part, connecting longwave and shortwave fragments of the monotonous neutral curves (see Section 2.2). The calculations results carried out for $D_s = 10^{-3}$; $B = 0.05$; $E = 0.01$; $S = 10^2$ are shown in Fig. 7.24. With the increase in A, all neutral curves displace to

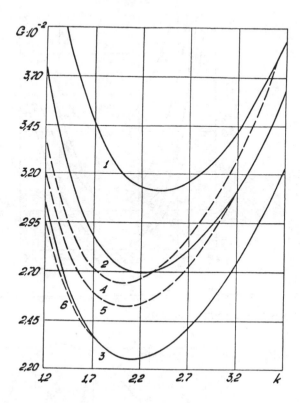

Figure 7.19. The neutral curves for an air-water system (heating from below); $Sc=10^3$; $A=1$ (lines 1,4); 6 (2,5); 31 (3,6).

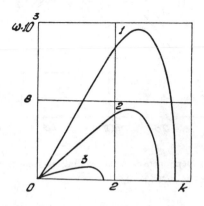

Figure 7.20. Dependences of the oscillations frequency on the wave number; $A=1$ (line 1); 6 (2); 31 (3).

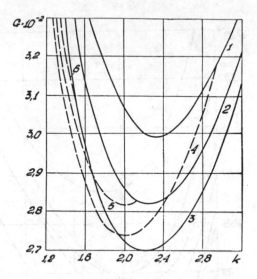

Figure 7.21. The neutral curves for an air-water system; $Sc=10^2$; $A=0.2$ (lines 1,4); 0.4 (2,5); 0.6 (3,6).

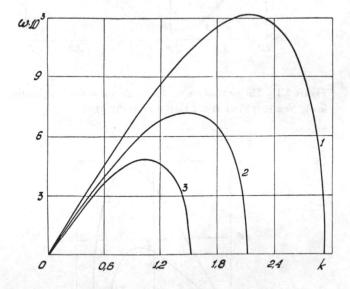

Figure 7.22. Dependences of the oscillations frequency on the wave number; $A=0.2$ (line 1); 0.4 (2); 0.6 (3).

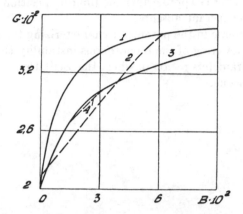

Figure 7.23. Dependences of the minimized according to wave number k Grashof number $G(B)$; $A=0$ (lines 1,2) 0.4 (3,4); $Sc=10^2$.

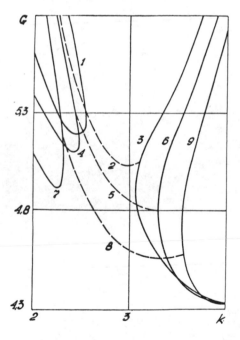

Figure 7.24. The neutral curves for a transformer oil-formic acid system; $A=0$ (lines 1–3); 1 (4–6); 10 (7–9).

the region of smaller G approaching the limiting position corresponding to the case of surfactant absence.

Thus, the increase in the parameter characterizing the surfactant solubility leads to lowering of the monotonous instability threshold and to narrowing of parameters region in which the oscillatory instability is the most dangerous one.

CHAPTER 8

Convective Stability of a System with Release of Heat on the Interface

Convective flow regimes can be essentially influenced by an additional heat source presence stipulated by heat release or heat absorption on the interface. The origin of surface heat source can be connected with a heterogeneous chemical reaction, fluid evaporation, emission, absorption and so on. In Section 8.1 thermogravitational convection generation in the presence of two temperature heterogeneity sources: solid boundaries temperature differences and heat-release or heat-absorption on the interface is studied. Heat release influence on thermocapillary convection in a two-layer system is considered in Sections 8.2, 8.3.

8.1. Thermogravitational convection with superficial heat release

Thermogravitational convection onset in a system of two infinite horizontal layers of immiscible fluids when heating from below was considered in Section 2.2. It was noted that stability loss can be stipulated both by monotonous and by oscillatory disturbances. Let us consider the surface heat release influence on convective stability of a two-layer system [61]. Let constant heat release of rate Q_0 ($Q_0 < 0$ relates to heat absorption) be set on the interface $z = 0$ supposed to be flat and non-deformable. Then except the source of the temperature heterogeneity connected with the solid boundaries temperature difference, an additional (surface) source comes into being.

Mechanical reference state is characterized by constant values of vertical temperature gradients A_m ($m = 1, 2$), found from heat balance condition on the interface $-\kappa_1 A_1 + \kappa_2 A_2 = Q_0$ and relation $A_1 a_1 + A_2 a_2 = -s\theta$ ($s = 1$ when heating from below, $s = -1$ when heating from above): $A_1 = -(s\theta\kappa_2 + Q_0 a_2)/(a_1\kappa_2 + a_2\kappa_1)$, $A_2 = -(s\theta\kappa_1 - Q_0 a_1)/(a_1\kappa_2 + a_2\kappa_1)$. Using notation of the second chapter we can formulate a boundary problem in the form (2.36), (2.37). Dimensionless temperature gradient at the steady state equals $A_1 = -(s + Qa\kappa)/(1 + \kappa a)$ in the upper fluid and $A_2 = -\kappa(s - Q)/(1 + \kappa a)$ in the lower fluid, where $Q = Q_0 a_1/\theta\kappa_1$. The

241

case $Q > 0$ corresponds to heat release and $Q < 0$ to heat absorption
on the interface. Equality $\lambda = 0$ determines infinite aggregate of neutral
curves $G = G(k, Q)$ related to different disturbances modes.

As shown in Section 2.2, at $Q = 0$ normal disturbances relative to
the given monotonous neutral curve are often localized mainly in one of
the fluids. In closing points of monotonous neutral curves relative to the
disturbances in different fluids the oscillatory instability can arise. Heat
release on the interface influences convection generation in the upper
and lower layers in different ways. With the increase in Q $(Q > 0)$, the
temperature gradient in the upper fluid increases and in the lower one
it decreases, changing the sign at $Q = 1$. Owing to this, the neutral
curves relative to the disturbances development in the upper (lower)
fluid displace to the region of smaller (larger) Grashof numbers. At
$Q < 0$ the neutral curves displacement taking place as $|Q|$ increases is
of the opposite character. As a result, the neutral curves can couple,
causing oscillatory instability.

1. Water-silicone oil system. Boundary problem was solved numerically by the Runge–Kutta method. Let us describe calculation results for a water-silicone oil N 200 system (see Table 1). Monotonous
neutral curves will be in solid lines and oscillatory curves will be in
dashed lines. In Fig. 8.1 two neutral curves having the lowest position
at $Q = 0$ relative to the monotonous generation of convection mainly
in the upper (curve 1) and lower (curve 2) fluids are presented. With
respect to the given above considerations, with the increase in Q, the
neutral curve 1 goes down and the curve 2 goes up. On the contrary,
at $Q < 0$ with the increase in $|Q|$ the neutral curves tend to meet. Pa-
rameter $|Q|$ increase makes the monotonous neutral curves close and
gives rise to a section of oscillatory instability in a longwave region (see
lines 3,4). At some value of $|Q|$ instability in a longwave region becomes
monotonous again and oscillatory instability is realized in the interval
$k_1 < k < k_2$. In a certain $|Q|$ range minimum value of the Grashof number is achieved on the oscillatory neutral curve, linking the longwave
and the shortwave fragments of monotonous neutral curves (lines 5–7).
Later on (see lines 8–10), monotonous disturbances again become the
most dangerous ones but now they are localized in the lower fluid.

Fig. 8.2 presents typical dependences of the neutral oscillations frequency on the wave number. In finite points of oscillatory neutral curves
on monotonous curves the oscillation frequency vanishes.

Thus, even in the systems for which instability has a monotonous
character in the absence of surface heat release, oscillatory instability
which under certain conditions can become the most dangerous one can
come into being.

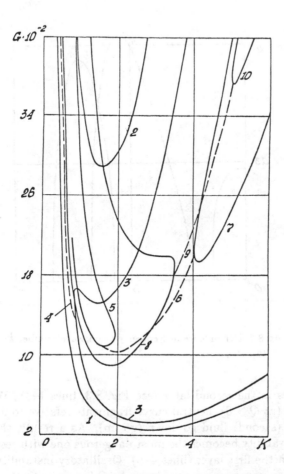

Figure 8.1. The neutral curves for a water-silicone oil system; $a=1.6$; $Q=0$ (lines 1,2); -0.8 (3,4); -3 (5–7); -3.2 (8–10).

2. Transformer oil-formic acid system. Let us consider now transformer oil-formic acid system (see Table 1) for which in the absence of surface heat sources or sinks there is a part of oscillatory neutral curve, linking longwave and shortwave fragments of the monotonous neutral curve. Let us note that disturbances in the second fluid correspond to the upper branch of the longwave fragment, and the lower branch of the shortwave fragment. Let us first describe the case $a = 0.667$ for which at $Q = 0$ the instability is connected with shortwave monotonous

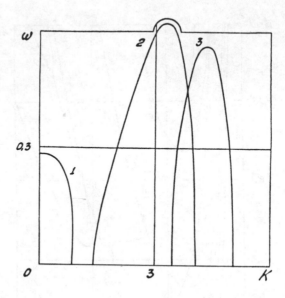

Figure 8.2. Dependence of frequency ω on wave number k; Q=-0.8 (line 1); -3 (2); -3.2 (3).

disturbances in the second layer (see Fig. 8.3, lines 1–3). With the increase in Q (at $Q > 0$), neutral curve fragments relative to disturbances in the first (second) fluid go down (go up). As a result, the longwave mode of instability becomes the most dangerous one with respect to disturbances in the first layer (lines 4–6). Oscillatory instability is not the most dangerous one at any Q values. As Q increases further, the oscillatory neutral curve is displaced to the shortwave region (lines 7,8). At $Q < 0$, with the increase in $|Q|$, the threshold of shortwave monotonous instability decreases and the longwave one increases; the interval of oscillatory instability existence displaces towards smaller wave numbers (lines 9–11).

Let now $a = 0.54$. At $Q = 0$ the longwave monotonous disturbances localized mainly in the upper fluid (see Fig. 8.4, lines 1–3) become the most dangerous ones. This type of disturbances is obviously the most dangerous one at $Q > 0$. At $Q < 0$ with the increase in $|Q|$ the monotonous neutral curve fragments relative to disturbances in the upper fluid are displaced towards larger G, and the fragments relative to disturbances in the lower fluid are displaced towards smaller G. As a

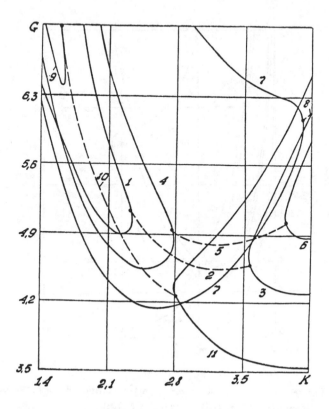

Figure 8.3. The neutral curves for a transformer oil-formic acid system; $a=0.667$; $Q=0$ (lines 1–3); 0.1 (4–6); 0.35 (7,8); -0.2 (9–11).

result, the neutral curves picture given in Fig. 8.4 (lines 4–6) is changed. Let us note that with the increase in $|Q|$ the region of oscillatory instability existence is getting narrower. At some $|Q|$ the neutral curves close causing their breakdown into two independet monotonous lines, and oscillations vanish (lines 7,8).

3. Water-mercury system. As shown in Section 2.3 some specific instability mechanism causing convection when heating from above acts parallel with the Rayleigh instability mechanism realized when heating from below in some fluid systems. Among such systems there is a water-mercury system at $10°$ C (see Table 2); $a = 1$. Heat release (heat absorption) being absent on the interface, temperature gradients in both the media are proportional ($A_2 = \kappa A_1$) and equally directed,

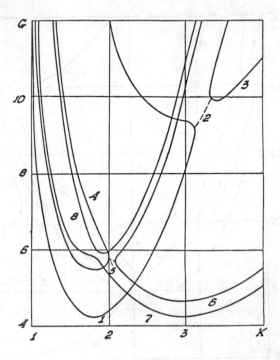

Figure 8.4. The neutral curves for a transformer oil-formic acid system; $a=0.54$; $Q=0$ (lines 1–3); -1 (4–6) -1.2 (7,8).

so depending on the heating method only one of the possible instability mechanisms becomes apparent. Surface heat sources being present, the signs of values A_m $(m = 1, 2)$ can turn out to be different.

Convection onset in systems with two different physical mechanisms of instability needs special analysis.

Let us introduce a new parameter:

$$G_Q = GQ = \frac{g\beta_1 Q_0 a_1^4}{\nu_1^2 \kappa_1},\qquad (8.1)$$

It determines heat release intensity on the interface irrelevant to value and sign of external heating. Using notation of (8.1) let us write down expression for dimensionless temperature gradients in the form

$$A_1 = -\frac{sG + a\kappa G_Q}{G(1 + \kappa a)}, \quad A_2 = -\frac{\kappa(sG - G_Q)}{G(1 + \kappa a)}. \qquad (8.2)$$

Let us discuss the surface heat release or heat absorption effect on the non-Rayleigh instability mode using the example of model system with $\beta \ll 1$, $\chi \ll 1$ discussed in 1 Section 2.3, before we pass to the analysis of numerical calculation results for a water-mercury system. Surface heat sources are easily seen to influence the convection threshold for this system only through temperature gradient A_1 in the medium with less temperature diffusivity. That is why heat release ($G_Q > 0$) causes instability stabilization when heating from above. On the contrary, heat absorption destabilizes the non-Rayleigh mode. At $G_Q < 0$ the system can turn out to be unstable even if there is no external heating; for its stabilization the intensive enough heating from below is necessary ($sG > 0$).

Let us go back to water-mercury system. If there are no surface heat sources ($G_Q = 0$), the neutral curve $sG(k)$ consists of two fragments. The first one, related to heating from below (see lines 1 in Fig. 8.5, 8.6) corresponds to Rayleigh convection in the water layer (weak induced motion is realized in mercury in this case). Mercury passivity can be explained by the fact that the local Rayleigh number for the water layer is 30 times larger than that for the mercury layer (see (2.39)). The second fragment (line 2 in Fig. 8.6) describes instability boundary for heating from above earlier given in Fig. 2.12 (line 1).

Let us first consider the case of heat release ($G_Q > 0$). According to (8.2) the temperature gradient in a water layer is directed downward at $sG > -a\kappa G_Q$ and increases with the increase in G_Q. That is why the threshold of Rayleigh convection generation in a water layer is natural to decrease with the increase in G_Q (see Fig. 8.5). According to numerical calculations at $G_Q = G_{Q*} \simeq 4300$ minimum value $sG(k)$ changes the sign. At $G_Q > G_{Q*}$ the neutral curves describe the Rayleigh instability stabilization when heating from above. Dependence of minimized according to k value $G_1 = \min_k sG(k)$ for the Rayleigh instability mode is shown in Fig. 8.7 (line 1).

As seen from the model system analysis, heating from above in the layer with low temperature diffusivity is necessary in order to realize the non-Rayleigh instability mechanism. That is why this instability mode is possible in the range $sG < -a\kappa G_Q$ in which in the upper fluid $A_1 > 0$. Change of maximum according to k value $G_2 = \max_k sG(k)$ with the increse in G_Q for the non-Rayleigh mode is given in Fig. 8.7

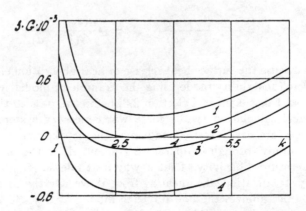

Figure 8.5. The neutral curves for Rayleigh instability with heat release; $G_Q = 0$ (line 1); 4300 (2); 7000 (3); 16000 (4).

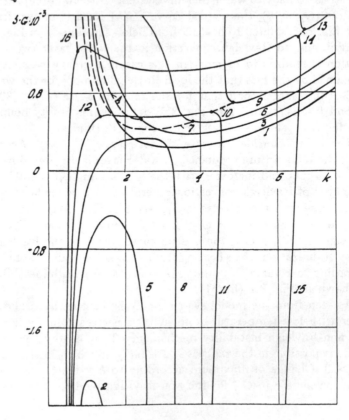

Figure 8.6. The neutral curves with heat absorption; $G_Q = 0$ (lines 1,2); -2500 (3–5); -4300 (6–8); -7000 (9–12); -16000 (13–16).

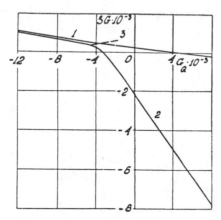

Figure 8.7. Boundaries of the mechanical stability region with respect to monotonous Rayleigh mode (line 1), non-Rayleigh mode (line 2) and oscillatory mode (line 3).

(line 2). In the region between lines 1 and 2 at $G_Q > 0$ the mechanical steady state is stable.

Let us now turn to the case of heat absorption ($G_Q < 0$) analysis. First, we shall discuss the evolution of neutral curves relative to Rayleigh instability. As $s = 1$ with the increase in $|G_Q|$ the temperature gradient in the upper layer decreases and in the lower one increases. As mentioned in this section with the reference to other fluid systems, neutral curves relative to convection generation in water are displaced towards sG increase and neutral curves for convection in mercury are displaced towards sG decrease. At parameters values for which the local Rayleigh numbers calculated according to temperature gradient values are of the same order for both media, oscillatory instability comes into being. As seen in Fig. 8.6, 8.8 the oscillatory neutral curve for the considered system appears in the wave numbers region, the numbers being smaller than the threshold wave number for a monotonous neutral curve. $|G_Q|$ still growing, oscillatory instability becomes more dangerous than monotonous one (see lines 6,7 Fig. 8.6). Minimized according to k value $G_3 = \min_k sG(k)$ for the oscillatory instability mode is given in Fig. 8.7 (line 3).

For the non-Rayleigh instability mode ($s = -1$) the increase in $|G_Q|$ causes an increase in positive temperature gradient in the upper fluid and

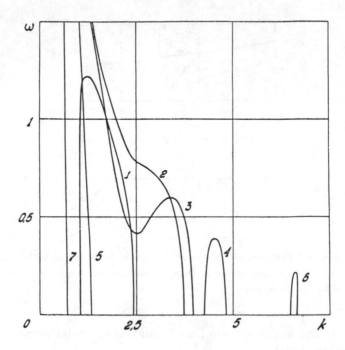

Figure 8.8. Dependence of frequency ω on wave number k; G_Q=-2500 (line 1); -4300 (2); -4800 (3); -7000 (4,5); -16000 (6,7).

sign change of the temperature gradient in the lower fluid (as $|G_Q| > G$). It was mentioned that to realize the instability mechanism considered, the temperature gradient must be positive only in the layer with the lower temperature diffusivity. That is why the non-Rayleigh neutral curve is kept at $G_Q < 0$ and its extreme G_2 increases with the increase in $|G_Q|$. As $|G_Q| > 3500$, value G_2 becomes positive. In this case mechanical steady state is kept only in some interval of parameter sG positive values; as sG increases, oscillatory instability develops, and as sG decreases, monotonous instability develops.

As $|G_Q|$ increases further, neutral curves related to oscillatory instability at $G > G_3$ and to monotonous instability at $G < G_2$ tend to meet. This approach causes first "a gap" on the graph scale $\omega(k)$ (see line 3 in Fig. 8.8) and then the oscillatory neutral curve fractures into two parts and couples with the monotonous neutral curve caused by heating from above (see lines 9–12 Fig. 8.6, lines 4,5 Fig. 8.8). In this case the region of the reference state stability disappears. As $|G_Q|$ increases further, parts of oscillatory neutral curves grow narrower (lines 14, 16 Fig. 8.6, lines 6,7 Fig. 8.8).

8.2. Thermocapillary convection with constant heat release

Let us now assume that the Archimedes force is negligibly small ($G = 0$). It has been mentioned in Sections 2.1, 2.5 that thermocapillary instability can be connected both with the monotonous and the oscillatory disturbances, the oscillatory instability being the only possible one under certain conditions.

Let us investigate the effect of surface heat source on thermocapillary instability of a two-layer system. In this paragraph surface heat release Q_0 is supposed to be of constant value; the case of heat extraction determined by temperature on the interface is considered in the next section. Boundary value problem (2.53)–(2.55) includes now equilibrium temperature gradients:$A_1 = -(s + Qa\kappa)/(1 + \kappa a)$ and $A_2 = -\kappa(s - Q)/(1 + \kappa a)$; (parameter Q is introduced in §8.1).

For the case of monotonous instability the problem admits analytical solution obtained in [44]. The expression for the critical Mr number in our notation has the form:

$$Mr(k) = 8k^2 \frac{1 + \kappa a}{P\kappa} \frac{(\kappa D_1 + D_2)(\eta B_1 + B_2)}{[s(\chi E_2 - E_1) - Q(\chi E_2 + a\kappa E_1)]} \qquad (8.3)$$

(compare with formula (2.57)).

To analyze the surface heat release effect on monotonous stability it is conveniently to use, instead of Q, the parameter

$$Mr_Q = MrQ = \frac{\alpha Q_0 a_1^2}{\eta_2 \nu_1 \kappa_1}. \qquad (8.4)$$

In contrast to Q, parameter Mr_Q does not depend on θ and is constant at temperature differences change between the upper and the lower system boundaries. In new variables (8.3) takes the form

$$Mr(k) = s \frac{8k^2(1 + \kappa a)(P\kappa)^{-1}(\kappa D_1 + D_2)(\eta B_1 + B_2) + Mr_Q(\chi E_2 + a\kappa)}{(\chi E_2 - E_1)} \qquad (8.5)$$

One can see that heat release on the interface ($Mr_Q > 0$) always stabilizes and heat absorption ($Mr_Q < 0$) always destabilizes the monotonous mode of instability. This effect becomes understandable from qualitative considerations. Hot spot arising on the interface causes a fluid influx from the solid boundaries side, and its spreading along the interface. If the interface is heated with respect to solid boundaries, then the arising influx of colder fluid will cause the temperature disturbance damping. If the interface is cooled, then the influx of warmer fluid will reinforce the temperature disturbance.

Let us consider a special case $\chi = 1$, $a = 1$. It was mentioned in Section 2.5 that monotonous instability at given parameters is not realized in the absence of heat release. Formula (8.5) shows that the boundary of monotonous instability has got in this case the form

$$Mr_Q = -\frac{8k^2(1+\kappa)(1+\eta)}{P\kappa}\frac{s_1 c_1 - k}{s_1^2 t_1 - k^3} \tag{8.6}$$

and does not depend on parameter Mr.

To get the boundaries of oscillatory instability the problem must be solved numerically.

Let us consider the system with parameters $\eta = \nu = 0.5$; $\kappa = \chi = P = a = 1$. We shall limit our consideration when heating from below. Monotonous instability takes place at $Mr_Q < Mr_{Q*} < 0$, where Mr_{Q*} is found from the extreme of expression (8.6). If $Mr_Q > Mr_{Q*}$ (particularly in the absence of heat absorption) oscillatory instability is the only possible mechanism of instability. Let us describe the results of neutral curve calculations carried out for fixed parameter Q (see Fig. 8.9). At $Q > 0$ with the increase in Q oscillatory neutral curve is stabilized; the monotonous neutral curve does not come into being. On the contrary, in the case $Q < 0$ with the increase in $|Q|$ the oscillatory neutral curve is displaced towards the region of smaller Mr numbers. Besides, at $|Mr_Q| = Mr|Q| > |Mr_{Q*}|$ the monotonous neutral curve arises. As parameter $|Q|$ increases ($Q < 0$), the monotonous mode of instability is destabilized more than the oscillatory one, and as a result, the monotonous instability becomes more dangerous. Graphs of the oscillation frequency ω depenence on the wave number k are given in Fig. 8.10. Fig. 8.11 presents the graph of the minimized according to k Mr_* number value dependence on parameter Mr_Q for the oscillatory (line 1) and monotonous (line 2) instability modes. In the region $Mr_Q > Mr_{Q*}$ oscillatory disturbances turn out to be the most dangerous ones and in the region $Mr_Q < Mr_{Q*}$ the same can be said about monotonous disturbances.

Let us consider now the system of real fluids: transformer oil-formic acid parameters of which are given in Table 1. Heat release being absent

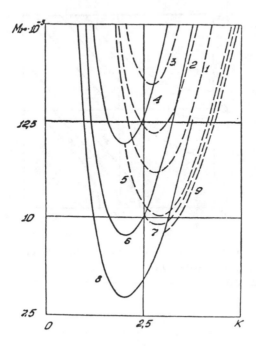

Figure 8.9. The neutral curves for the values of parameter Q=0 (line 1); 0.015 (2); 0.03 (3); -0.02 (4,5); -0.025 (6,7); -0.03 (8,9).

$(Q = 0)$, this system as it has been mentioned in Section 2.5, turns out to be unstable with respect to monotonous disturbances when heating from both the first and the second fluids side (see Fig. 8.12, line 1). Besides, oscillatory instability may arise in the longwave region (line 2). The conclusion about the stabilizing effect of heat release turns out to be fair not only for the monotonous but also for the oscillatory disturbances. At $Q > 0$ with the increase in Q all neutral curve fragments are stabilized (see Fig. 8.12, lines 3–6). Line 3 fragment for $s = 1$ is not presented because it is too close to line 1 on graph scale. At $Q < 0$ destabilization takes place (see lines 7–12). In the whole investigated region of parameter Q the neutral curve minimum is realized on monotonous disturbances. Dependence of ω on k for oscillatory disturbances is presented in the insert to Fig. 8.12. Numeration of lines in the insert corresponds to that of lines in the main graph.

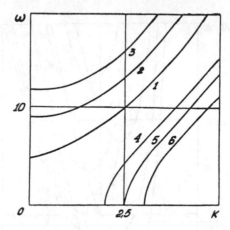

Figure 8.10. Dependence of oscillations frequency ω on wave number k for the values of parameter $Q=0$ (line 1) 0.015 (2); 0.03 (3); -0.02 (4); -0.025 (5); -0.03 (6).

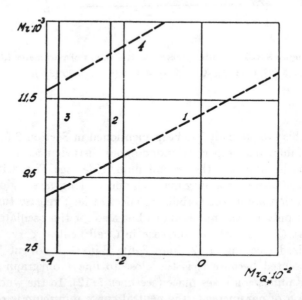

Figure 8.11. Boundaries of the stability regions with respect to monotonous and oscillatory disturbances with constant heat release (lines 1,2) and in the presence of heat release dependence on temperature ($Q_T=-2.4$; lines 3,4).

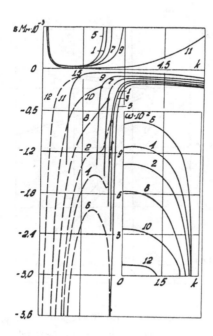

Figure 8.12. The neutral curves for a transformer oil-formic acid system; a=1.667; Q=0 (lines 1,2); 0.015 (3,4); 0.03 (5,6); -0.03 (7,8); -0.09 (9,10); -0.3 (11,12); in insert — dependence $\omega(k)$.

Fig. 8.13 presents stability region boundaries with respect to monotonous (line 1) and oscillatory (line 2) disturbances obtained from neutral curves extremes. Relative instability type is realized in the region placed to the left from the given in the figure boundaries. Monotonous disturbances are seen to be the most dangerous ones for this fluid system.

8.3. Superficial temperature-dependent heat release

Let us turn to investigation of the case when the surface heat release Q_Γ is not constant but is determined by boundary temperature T_Γ (it takes place, for example, in the case of a heterogeneous chemical reaction): $Q_\Gamma = Q_\Gamma(T_\Gamma)$. In this case heat release value Q_0, realized under mechanical equilibrium conditions, is calculated from the heat balance condition; heat produced on the interface equal to heat transfered to

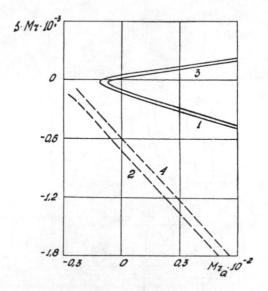

Figure 8.13. Boundaries of the stability regions with respect to monotonous and oscillatory disturbances with constant heat release (lines 1,2) and in the presence of heat release dependence on temperature (Q_T=1.01; lines 3,4).

the system's solid boundaries. Let the temperatures of the upper and the lower plates be T_1 and T_2. Using the expression for temperature gradients in fluids A_m ($m = 1, 2$) given in Section 8.1, we find that equilibrium temperature values of the boundary T_0 and heat release Q_0 are determined from equations

$$T_0 = \frac{T_1 \kappa_1 a_2 + T_2 \kappa_2 a_2 + Q_0 a_1 a_2}{a_1 \kappa_2 + a_2 \kappa_1}, \qquad (8.7)$$

$$Q_0 = Q_\Gamma(T_0). \qquad (8.8)$$

Fig. 8.14 presents different opportunities of functions (8.7) (lines 1–3) and (8.8) (line 4) graphs disposition. The system is seen to have several solutions, a unique solution and no solutions. Later on, at least one solution of system (8.7), (8.8) is supposed to exist.

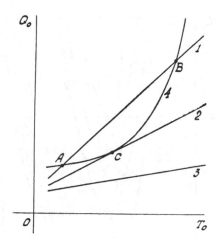

Figure 8.14. Examples of the system solutions graph construction (8.5). Line 1 — the case of two solutions (points A and B), line 2 — of one solution (point C), line 3 — absence of solutions.

Heat release dependence on temperature leads to the change of boundary condition for heat flow disturbances on the interface which takes the form:

$$z = 0: \quad \kappa T_1' = T_2' - Q_T \cdot T_1,$$

where

$$Q_T = \left(\frac{dQ_\Gamma}{dT_\Gamma}\right)_{T_\Gamma = T_0} \cdot \frac{a_1}{\kappa_2}.$$

The rest of the equations and boundary conditions keep the form (2.53)–(2.55).

Monotonous instability boundary $\lambda = 0$ is found analytically:

$$Mr(k) =$$
$$_s\frac{8(1 + \kappa a)(P\kappa)^{-1}k[k(\kappa D_1 + D_2 - Q_T)](\eta B_1 + B_2) + Mr_Q(\chi E_2 + a\kappa E_1)}{\chi E_2 - E_1}$$
$$(8.9)$$

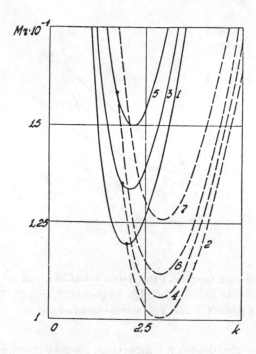

Figure 8.15. The neutral curves as Q=-0.02; Q_T=0 (lines 1, 2); -0.5 (3,4); -1.1 (5,6); -2.4 (7).

It is seen from expression (8.9) that in case heat release increases with the increase in temperature ($Q_T > 0$), the instability threshold decreases; and with release decreasing, instability threshold increases. The problem statement for thermocapillary convection makes sense only in the range

$$Q_T < Q_{T*} = \kappa + \frac{1}{a}; \tag{8.10}$$

at $Q_T > Q_{T*}$ the steady state is unstable with respect to longwave disturbances ($k \to 0$) independent of parameters Mr, Mr_Q values which meets heat explosion effect [44]. It is easy to prove that condition (8.10) is fulfilled for the steady state relative to point A in Fig. 8.14 and is not fulfilled for point B. Equality $Q_T = Q_{T*}$ takes place for point C.

At $\chi = 1$, $a = 1$ the boundary of monotonous instability has the form

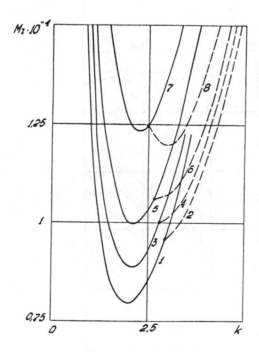

Figure 8.16. The neutral curves as Q=-0.03: Q_T=0 (lines 1,2); -0.5 (3,4); -1.1 (5,6); -2.4 (7,8).

$$Mr_Q(k) = -\frac{8(1+\kappa)(1+\eta)k[k(\kappa+1)C_1 - Q_T S_1](S_1 C_1 - k)}{\kappa P(S_1^3 - k^3 C_1)} \quad (8.11)$$

and does not depend on Mr.

Let us describe value Q_T effect on stability of the systems considered in Section 8.2. Examples of neutral curves for the system with parameters $\eta = \nu = 0.5$; $\kappa = \chi = P = a = 1$ when heating from below constructed at values $Q < 0$, $Q_T \leq 0$ are given in Fig. 8.15 and 8.16. Both monotonous and oscillatory instability modes stabilize with the increase in $|Q_T|$. Stability region boundaries obtained from the calculation of neutral curves extremes are presented in Fig. 8.11. Monotonous instability is realized at $Mr_Q < Mr_{Q*} < 0$, where Mr_{Q*} is extremum of (8.11). At $Mr_Q > Mr_{Q*}$, only oscillatory instability is possible. As

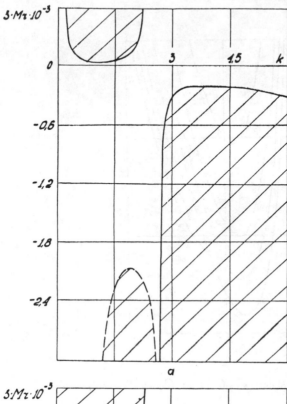

a

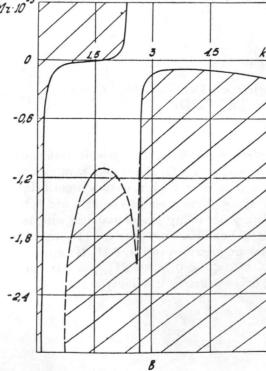

в

Figure 8.17(a,b). The regions of disturbances increase (dashed) and damping as $Q=0.03$: *a*) $Q_T=0$; *b*) $Q_T=1.1$.

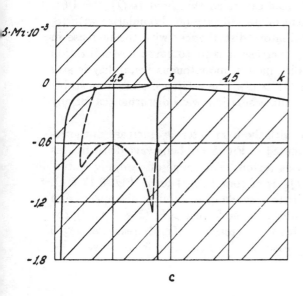

c

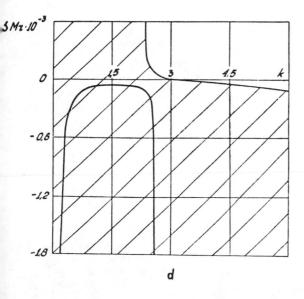

d

Figure 8.17(c,d). The regions
of disturbances increase (dashed)
and damping as $Q=0.03$: c)
$Q_T=3.6$; d) $Q_T=4.8$.

$|Q_T|$ grows, the oscillatory instability region grows wider owing to the decrease in Mr_{Q_*}.

We shall consider now a system of two real fluids: transformer oil-formic acid. At $a = 1.667$ heat explosion threshold is $Q_{T_*} = 1.01$. Fig. 8.13 presents displacement of stability regions boundaries with respect to monotonous and oscillatory disturbances with the increase in Q_T from 0 till $Q_{T_*} - 0$. Q_T increase is seen to cause destabilization with respect to both instability modes, monotonous instability is still the most dangerous one. At $Q_T > Q_{T_*}$ the reference state is unstable with respect to monotonously increasing longwave disturbances for any sMr, Mr_Q.

The neutral curves change with the parameter Q_T increase in the region of heat explosion is presented in Fig. 8.17. Longwave disturbances $(k \to 0)$ always increase in this region. The range of damping disturbances at not large enough Q_T is of complicated form (Fig. 8.17 b, c). At large enough Q_T the oscillatory mode of instability disappears and the disturbances damping region gets the form $k > k_*$ (Fig. 8.17 d); the boundary wave number k_* with the increase in Mr, decreases as $s > 0$ (heating from below) and increases as $s < 0$ (heating from above).

REFERENCES

1. Abritska M.Yu., Pagodkina I.E., Mikelson A.E. MHD — convection in a two-layer system when heating from below // Magnetic hydrodynamics. 1983. No. 4. P. 94–98.
2. Adilov R.S., Putin G.F., Shaidurov G.F. Convective stability of two immiscible fluids in a horizontal layer // Uch. zap. Perm univ. 1976. N 362. Hydrodynamics. Iss. 8. P. 16-20.
3. Alekseev V.V., Aleksandrov A.A. Two-dimensional convection model in a two-layer fluid // Izv. AN SSSR. FAO. 1973. V. 9 No. 8. P. 837-850.
4. Alekseev V.V., Gusev A.M. Free convection in geophysical processes // UFN. 1983. V. 141. Iss. 2. P. 311-342.
5. Antanovskii L.K., Kopbosynov B.K. Nonsteady thermocapillary drift of a drop of viscous fluid // Prikl. Mekh i Tekhn. Fiz. 1986. No. 2. P. 59-64.
6. Arnold V.I. Additional chapters of ordinary differential equations // M.: Nauka. 1978. 304 P.
7. Avduevskii V.S., Barmin I.V., Grishin S.D. Space production problems // M.: Mashinbuilding. 1980. 221 P.
8. Babskii B.G., Sklovskaya I.L. Hydrodynamics in weak power fields. Onset of steady thermocapillary convection in a sphere fluid layer under microgravity conditions // Izv. AN SSSR. Mekh. Zhidk. i Gaza. 1969. No. 3. P. 92–99.
9. Badratinova L.G. Dynamic and heat instability of free fluid boundaries in a weak power fields // Avtoref. cand. diss. Novosibirsk. 1983. 16 P.
10. Badratinova L.G. Nonlinear equations of long waves in the problem of thermocapillary convection in a two-layer fluid // Continious medium dynamics. Novosibirsk. 1985. Iss. 69. P. 3–18.
11. Bauer H.F., Eidel W. Marangoni convection in a spherical liquid system // Acta Astronautica. 1987. V.15. No. 5. P. 275–290.
12. Berezovskii E.I., Perelman T.L., Romashko E.A. On convective stability in the system of two infinite horizontal layers of immiscible fluids // Inzh. Fiz. Zh. 1974. V.27. No. 6. P. 1098–1108.
13. Berg J.C., Acrivos A. The effect of surface active agents on convection cells induced by surface tension // Chem. Eng. Sci. 1965. V. 20. P. 737-745.
14. Birger B.I. Viscosity jump influence on convective stability of the upper Earth mantle // Izv. AN SSSR. Fiz. Zemli. 1976. No. 8. P. 8-15.

264

15. Birger B.I., Shlesberg S.G. Convective instability of two-layer models of Earth mantle // Izv. AN SSSR. Fiz. Zemli. 1976. No. 9. P. 3–14.
16. Birikh R.V. On thermocapillary convection in a horizontal fluid layer // Prikl. Mekh. i Tekhn. Fiz. 1966. No. 3. P. 67–72.
17. Bratukhin Yu.K. Thermocapillary drift of a viscous fluid drop // Izv. AN SSSR Mekh. Zhidk. i Gaza. 1975. No. 5. P. 156–161.
18. Bratukhin Yu.K. Overflow of a gas bubble by the flow of heterogeneous fluid at small Marangoni numbers // Inzh. Fiz. Zh. 1977. V. 32. No. 2 P. 251–256.
19. Bratukhin Yu.K. Briskman V.A., Zuev A.L., Pshenichnikov A.F., Rivkind V.Ya. Experimental investigation of thermocapillary drift of gas bubbles in fluid // Hydrodynamics and heat-mass transfer in microgravity. M.: Nauka. 1982. P. 98–109.
20. Bratukhin Yu.K., Evdokimova O.A., Pshenichnikov A.F. Gas bubbles motion in heterogeneously heated fluid // Izv. AN SSSR. Mekh. Zhidk. i Gaza. 1979. No. 5. P. 55–57.
21. Brian P.L. Effect of Gibbs adsorption on Marangoni instability // AIChE J. 1971. V. 17. No. 4. P. 765–772.
22. Burde G.I., Simanovskii I.B. Determinition of critical numbers spectrum of a two-layer system by the finite difference method // Heat convection and heat transfer investigation. Sverdlovsk. UNC AN SSSR. 1981. P. 58–63.
23. Burde G.I., Simanovskii I.B. Determination of convective stability boundaries of a two-layer system // Prikl. Math. i Mekh. 1979. V. 43. No. 6. P. 1008–1013.
24. Busse F.H. The stability of the finite-amplitude cellular convection and its relation to an extremum principle // J. Fluid Mech. 1967. V. 30. Pt. 4. P. 625–641.
25. Busse F.H. A model of time-periodic mantle flow // Geophys. J. R. Astr. Soc. 1978. V. 52. P. 1–12.
26. Busse F.H. On the aspect rations of two-layer mantle convection // Phys. Earth Planet. Inter. 1981. V. 24. P. 320-324.
27. Carpenter B., Homsy G.M. The effect of surface contamination on thermocapillary flow in a two-dimensional slot. Pt. 2. Partially contaminated interfaces // J. Fluid Mech. 1985. V. 155. P. 429–439.
28. Cerisier P., Occelli R., Pantaloni J. Velocity and temperature fields in Benard-Marangoni instability // Physicochemical Hydrodynamic. 1986. V. 7. No. 4. P. 191–206.
29. Chernatynskii V.I., Shliomis M.I. Convection close to the critical Rayleigh numbers at nearly vertical temperature gradient// Izv. AN SSSR. Mekh. Zhidk. i Gaza. 1973. No. 1. P. 64–70.

30. Chu X.-L., Velarde M.G. Free surface convection in a bounded cylindrical geometry: note on the role of surface adsorption and solute accumulation at the air-liquid interface // Int. J. Heat Mass Transfer. 1984. V. 31. No. 10. P. 1979–1982.

31. Cloot A., Lebon G. A nonlinear stability analysis of the Benard-Marangoni problem // J. Fluid. Mech. 1984. V. 145. P. 447-469.

32. Cserpes L., Rabinovicz M. Gravity and convection in a two-layer mantle // Earth and Planetary Science Letters. 1985/86. V. 6. P. 193–207.

33. Dalle Vedove W., Bisch P.M., Sanfeld A. Interfacial hydrodynamic instability induced by a stable surface chemical reactons // J. Non-Equilib. Thermodyn. 1980. V. 5. P. 35–53.

34. Daly B.J. Numerical study of a two-fluid Rayleigh-Taylor instability // Phys. Fluids. 1967. V. 10. No. 2. P. 297–307.

35. Daly B.J. Numerical study of the effect of surface tension on the interface instability // Phys. Fluid. 1969. V. 12. No. 7. P. 1340–1354.

36. Dzhoseff D. Fluid motion stability // M.: Mir. 1981. 640 P.

37. Earnshaw J.C. Surface viscosity of water // Nature. 1981. V. 292. No. 5819. P. 138–139.

38. Elenin G.G., Kalachinskaya I.S., Solomatin S.V. On Marngoni instability in a gas-fluid system // Differential equations. 1985. V. 21. No. 7. P. 1171–1179.

39. Ellsworth K., Schubert G. Numerical models of thermally and mechanically coupled two-layer convection of highly viscous fluids // Geophys. Journal. 1988. V. 93. No. 2. P. 347–363.

40. Elsasser W.M. Two-layer model of upper-mantle circulation // J. Geoph. Res. 1971. V. 76. No. 20. P. 4744–4753.

41. Faddeev D.E., Faddeeva V.N. Calculation methods of linear algebra // M.: Fizmathgiz. 1960. 734 P.

42. Ferm E.N., Wollkind D.J. Onset of the Rayleigh-Benard-Marangoni instability; comparison between theory and experiment // J. Non-Equilib. Thermodyn. 1982. V. 7. No. 3. P. 170–190.

43. Free convection in an atmosphere and in a ocean // Edited by A.M. Gusev. M.: Moscow university. 1979. 139 P.

44. Frencel G., Linde H. Linear analysis of Marangoni instability in a two-phase system with heat source or run-off on interphase boundary // Theor. Osn. Khim. Tekhnol. 1986. V. 20. No. 1. P. 28–36.

45. Fujinawa K., Nozawa M., Imaishi N. Effects of desorption and absorption of surface tension-lowering solutes on liquid-phase mass transfer coefficients at a turbulent gas-liquid interface // J. Chem. Eng. Jap. 1978. V. 11. No. 2. P. 107–111.

46. Garcia-Ybarra P.L., Velarde M.G. Oscillatory Marangoni-Benard interfacial instability and capillary-gravity waves in single-and two-component liquid layers with or without Soret thermal diffusion // Phys. Fluids. 1987. V. 30. No. 6. P. 1649–1655.

47. Gershuni G.Z., Zhukhovitsky E.M. Convective stability of two immiscible fluids in a spherical vessel // Uch. zap. Perm univ. 1968. No. 184. Hydrodynamics. Iss 1. P. 57–73.

48. Gershuni G.Z., Zhukhovitsky E.M. Convective stability of incompressible fluid // M.: Nauka. 1972. 392 P.

49. Gershuni G.Z., Zhukhovitsky E.M. On the instability of a horizontal layers system of immiscible fluids when heating from above // Izv. AN SSSR. Mekh Zhidk. i Gaza. 1980. No. 6. P. 28–34.

50. Gershuni G.Z., Zhukhovitsky E.M. On monotonous and oscillatory instability of a two-layer immiscible fluids system heated from below // Dokl. AN SSSR. 1982. V. 265. No. 2. P. 302–305.

51. Gershuni G.Z., Zhukhovitsky E.M. On convecive instability of a two-layer system with heat-insulated boundaries // Izv. AN SSSR. 1986. No. 2. P. 22-28.

52. Gershuni G.Z., Zhukhovitsky E.M., Nepomnyashchy A.A., Simanovskii I.B. Development of disturbances and transition to turbulence at the convection in two-layer systems // II IUTAM symposium on laminar-turbulent transition. Novosibirsk. 1984. Proceedings. Springer-Verlag. Berlin. Heidelberg. 1985. P. 749–754.

53. Gershuni G.Z., Zhukhovitsky E.M., Nepomnyashchy A.A., Simanovskii I.B. Convective stability of a two-layer system // Fifth. National Congress on Theoretical and Applied Mechanics. Varna. 1985. Proceedings. V. 1. Sofia. 1985. P. 211–216.

54. Gershuni G.Z., Zhukhovitsky E.M., Pershina E.A. On convective onset in some two-layer systems // Convective flows. Perm. 1983. P.25–31.

55. Gershuni G.Z., Zhukhovitsky E.M., Simanovskii I.B. On stability and finite-amplitude motions in a two-layer system heated from above // Convective flows. Perm. P. 3–11.

56. Gershuni G.Z., Zhukhovitsky E.M., Tarunin E.L. Secondary steady convective motions in a flat vertical fluid layer // Izv. AN SSSR. Mekh. Zhidk. i Gaza. 1968. No. 5. P. 130–136.

57. Gilev A.Yu., Nepomnyashchy A.A., Simanovskii I.B. Generation of thermocapillary and thermogravitational convection in an air-water system // Nonisothermal flows of viscous fluid. Sverdlovsk. UNC AN SSSR. 1985. P. 24–27.

58. Gilev A.Yu., Nepomnyashchy A.A., Simanovskii I.B. Generation of thermogravitational convection in a two-layer system in the presence

of surfactant on the interface // Prikl. Mekh. i Tekhn. Fiz. 1986. No. 5. P. 76–81.

59. Gilev A.Yu., Nepomnyashchy A.A., Simanovskii I.B. Onset of oscillatory thermogravitational convection in a two-layer system when heating from below // Dynamics of viscous fluid. Sverdlovsk. UNC AN SSSR. 1987. P. 36–37.

60. Gilev A.Yu., Nepomnyashchy A.A., Simanovskii I.B. Onset of convection in a two-layer system stipulated by a combined influence of Rayleigh and thermocapillary instability mechanisms // Izv. AN SSSR. Mekh. Zhidk. i Gaza. 1987. No. 1. P. 166–170.

61. Gilev A.Yu., Nepomnyashchy A.A., Simanovskii I.B. Thermogravitational convection in a two-layer system with heat release on the interface // Izv. AN SSSR. Mekh. Zhidk. i Gaza. 1990. No. 1. P. 173–175.

62. Gilev A.Yu., Simanovskii I.B. Numerical investigation of termogravitational convection in a two-layer system // Convective flows. Perm. 1985. P. 10–14.

63. Gilev A.Yu., Simanovskii I.B. Finite-amplitude thermocapillary convecton in a two-layer system // Inzh. Fiz. Zh. 1987. V. 52. No. 2. P. 244–247.

64. Golitsin G.S. Investigation of convection with geophysical applications and analogies // L.: Hydrometheoizdat. 1980. 55 P.

65. Golovin A.A., Gupalo Yu. P., Ryazantsev Yu.S. On chemothermocapillary effect for a drop motion in a fluid // Dokl. AN SSSR. 1986. V. 290. No. 1. P. 35–39.

66. Goodrich F.C. The theory of capillary excess viscosities // Proceeding of Royal Society. London. 1981. A-374. P. 341–370.

67. Gumerman R.J., Homsy G.M. Convective instabilities in concurrent two-phase flow. Pt. I. Linear stability // AIChE J. 1974. V. 20. No. 5. P. 981–988.

68. Gumerman R.J., Homsy G.M. Convective instabilities in concurrent two-phase flow. Pt. II. Global stability // AIChE J. 1974. V. 20. No. 6. P. 1161–1167.

69. Gumerman R.J., Homsy G.M. Convective instabilities in concurrent two-phase flow. Pt. III. Experiments // AIChE J. 1974. V. 20. No. 6. P. 1167–1172.

70. Harlow F.H. The particle-in-cell computing method for fluid dynamics // Methods in Computational Physics. V. 3. Fundamental Methods in Hydrodynamics. 1964. 319 P.

71. Harlow F.H., Welch J.E. Numerical calculations of time-dependent viscous incompressible flow of fluid with free surface // Phys. Fluids. 1965. V. 8. No. 12. P. 2182–2189.

72. Hennenberg M., Bisch P.M., Vignes-Adler M., Sanfeld A. Mass transfer, Marangoni effect and instability of interfacial longitudinal waves. I. Diffusional exchanges // J. Colloid and Interface Sci. 1979. V. 69. No. 1. P. 128–137.

73. Hennenberg M., Bisch P.M., Vignes-Adler M., Sanfeld A. Mass transfer, Marangoni effect and instability of interfacial longitudinal waves // J. Colloid and Interface Sci. 1980. V. 74. No. 2. P. 495–508.

74. Hennenberg M., Sanfeld A., Bisch P.M. Adsorption-desorption barrier, diffusional exchanges and surface instabilities of longitudinal waves for aperiodic regimes // AIChE J. 1981. V. 27. No. 6. P. 1002–1008.

75. Hennenberg M., Bish P.M., Vin-Adler M., Sanfeld A. Interface instability and longitudinal waves in a fluid-fluid system // Hydrodynamics of interphase surfaces. M.: Mir. 1984. 210 P.

76. Hydrodynamics and heat-mass exchange in microgravity / edited by V.S. Avduevskii, V.I. Polezhaev // M.: Nauka. 1982. 264 P.

77. Hydrodynamics and transfer processes in microgravity // Sverdlovsk. UNC AN SSSR. 1983. 168 P.

78. Hydrodynamics of interphases surfaces // Transl. from Engl. Edited by Yu.A. Buevich and L.M. Rabinovich. M.: Mir. 1984. 210 P.

79. Hydrodynamics of microgravity / Edited by A.D. Myshkis // M.: Nauka. 1976. 504 P.

80. Imaishi N., Fujinawa K. Theoretical study of the stability of two-fluid layers // J. Chem. Eng. Jap. 1974. V. 7. No. 2. P. 81–86.

81. Imaishi N.,Fujinawa K. Thermal instability in two-fluid layers // J. Chem. Eng. Jap. 1974. V. 7. No. 2. P. 87–92.

82. Ivanova S.V., Popel A.S. On the motion of a viscous fluid drop by the action of insolubal surface-active agent // Izv. AN SSSR. Mekh. Zhidk. i Gaza. 1974. No. 2. P. 63–68.

83. Kazaryan T.S., Yakovleva A.A. Transportation and storage of oil and hydrocarbon materials // VNIIONG. 1975. No. 8. P. 5–7.

84. Khan W. Eddy damping by surface active agents in interfacial turbulence // Math. Modell. 1985. V. 6. No. 2. P. 97–109.

85. Knight R.W., Palmer M.E. Simulation of free convection in multiple fluid layers in an enclosure by finite differences // Numerical properties and methodologies in heat transfer. Hemisphere. Washington. 1983. P. 305–319.

86. Koch M., Yuen D.A. Surface deformation and geoid anomalies over single and double-layered convective systems // Geophys. Res. Lett. 1985. V. 12. No. 10. P. 701–704.

87. Koschmieder E.L. On convection under an air surface // J. Fluid Mech. 1967. V. 30. Pt. I. P. 9–15.

88. Koschmieder E.L., Biggerstaff M.I. Onset of surface-tension driven Benard convection // J. Fluid Mech. 1986. V. 167. P. 49–64.

89. Krylov V.S. Theoretical aspects of intensification of interphase exchange process // Theor. Osn. Khim. Tekhnol. 1983. V. 17. No. 1. P. 15–30.

90. Kuskova T.V., Chudov L.A. On approximate boundary conditions for vortex at calculation of the flows of viscous incompressible fluid // Computational methods and programming. VC Moscow State Univ. Iss. 11. 1968. P. 27–31.

91. Kuznetsov E.A., Spektor M.D. On weakly supercritical convection / Prikl. Mekh. i Tekhn. Fiz. 1980. No. 2. P. 76–86.

92. Landau L.D., Lifshits E.M. Hydrodynamics // M.: Nauka. 1986. 736 P.

93. Levich V.G. Physicochemical hydrodynamics // M.: AN SSSR. 1952. 538 P.

94. Levich V. The influence of surface-active substances on the motion of liquids // Physicochemical hydrodynamics. 1981. V. 2. No. 2/3. P. 85–94.

95. Limbourg-Fontaine M.C., Petre G., Legros J.C. Texus 8 Experiment: effects of surface tension minimum on thermocapillary convection // Physicochemical Hydrodynamics. 1985. V. 6. No. 3. P. 301–310.

96. Linde H., Shvarts P., Vilke H. Dissipative structures and nonlinear kinetics of Marangoni instability // Hydrodynamics of interphase surfaces. M.: Mir. 1984. P. 79–116.

97. Myrum T.A., Sparrow E.M., Prata A.T. Numerical solutions for natural convection in a complex enclosed space containing either air-liquid or liquid-liquid layers // Numerical Heat Transfer. 1986. V. 10. P. 19–43.

98. Napolitano L.G. Thermodynamics and dynamics of pure interfaces // Acta Astronautica. 1978. V. 5. No. 9. P. 655–670.

99. Napolitano L.G., Golia C., Viviani A. Numerical simulation of unsteady thermal Marangoni flows // Proc. of the 5th Eur. Symp. on Material Sciences under Microgravity. Schloss Elmau. 1984. ESA-SP-222. 1984. P. 251–258.

100. Napolitano L.G., Monti R., Russo G. Marangoni convection in one- and two-liquids floating zones // Naturwissenchaften. 1986. V. 73. P. 352–355.

101. Narayanan R. Observation on thermoconvection for bilayers in containers of arbitrary shape // J. Eng. Math. 1983. V. 17. No. 3. P. 223–238.

270

102. Nepomnyashchy A.A. On secondary space-periodical motions in unlimited space // Uch. zap. Perm univ. 1976. No. 152. Hydrodynamics. Iss. 9. P. 77–86.

103. Nepomnyashchy A.A., Simanovskii I.B. Thermocapillary convection
in a two-layer system // Dokl. AN SSR. 1983. V. 272. No. 4. P.
825–827.

104. Nepomnyashchy A.A., Simanovskii I.B. Thermocapillary convection
in a two-layer system // Izv. AN SSSR. Mekh. Zhidk. i Gaza. 1983.
No. 4. P. 158–163.

105. Nepomnyashchy A.A., Simanovskii I.B. Numerical investigation of
thermocapillry convection in a two-layer systems // Hydromechanics
and transfer process in microgravity. Sverdlovsk. UNC AN SSSR.
1983. P. 161–166.

106. Nepomnyashchy A.A., Simanovskii I.B. Numerical investigation of
convection in a system of two fluids with distorted interface // Numerical methods of viscous fluids dynamics. Proceedings of IX all
Union school. Novosibirsk. 1983. P. 246–250.

107. Nepomnyashchy A.A., Simanovskii I.B. On oscillatory regimes of convection in a two-layer system // Inzh. Fiz. Zh. 1984. V. 46. No.
5. P. 845–849.

108. Nepomnyashchy A.A., Simanovskii I.B. On convection onset in a two-
layer system // Hydrodynamic and convective stability of incompressible fluid. Sverdlovsk. UNC AN SSSR. 1984. P. 10–18.

109. Nepomnyashchy A.A., Simanovskii I.B. Thermocapillary and thermogravitational convection in a two-layer system with curved interface
// Izv. AN SSSR. Mekh. Zhidk. i Gaza. 1984. No. 3. P. 175–179.

110. Nepomnyashchy A.A., Simanovskii I.B. On oscillatory convective instability of two-layer systems in the presence of thermocapillay effect
//Prikl. Mekh. i Tekhn. Fiz. 1985. No. 1. P. 62–65.

111. Nepomnyashchy A.A., Simanovskii I.B. Thermocapillary convection
in two-layer systems in the presence of surfactant on the interface //
Izv. AN SSSR. Mekh. Zhidk. i Gaza. 1986. No. 2. P. 3–8.

112. Nepomnyashchy A.A., Simanovskii I.B. Onset of oscillatory thermocapillary convection in the systems with deformable interface // Izv.
AN SSSR. Mekh. Zhidk. i Gaza. 1991. No. 4. P. 11–16.

113. Nepomnyashchy A.A., Simanovskii I.B. Numerical investigation of
thermocapillary convection in a two-layer system // Cybernetics
problems. Calculating aerohydromechanics / edited by O.M. Belotserkovskii. M.: 1987. P. 63–71.

114. Nepomnyashchy A.A., Simanovskii I.B. Mechanisms of oscillatory
generation in systems with an interface between media // I Con-
ference on Mechanics. Proceedings. V. 8. Praha. 1987. P. 86–89.

115. Nepomnyashchy A.A., Simanovskii I.B. Convective stability of a two-layer system heated from above in the presence of surfactant on the interface // Convective flows. Perm. 1987. P. 3–6.

116. Nepomnyashchy A.A., Simanovskii I.B. Onset of thermocapillary convection in a two-layer system in the presence of soluble surfactant // Izv. AN SSSR. Mekh. Zhidk. i Gaza. 1988. No. 2. P. 187–191.

117. Nepomnyashchy A.A., Simanovskii I.B. Onset of thermogravitational convection in the presence of soluble surfactant on the interface // Prikl. Mekh. i Tekhn. Fiz. 1989. No. 1. P. 146–149.

118. Nepomnyashchy A.A., Simanovskii I.B. Onset of oscillatory convection in a two-layer system, stipulated by the presence of surfactant on the interface // Dokl. AN SSSR. 1989. V. 306. No. 2. 310–313.

119. Nield D.A. Surface tension and buoyancy effects in cellular convection // J. Fluid Mech. V. 19. P. 341–352.

120. Nitshke U., Shvarts P., Krylov V.S., Linde H. The influence of the induced cellular convection on heat- and masstransfer through the mooving interphase boundary // Theor. Osn. Khim. Tekhnol. 1985. V. 19. No. 3. P. 311–316.

121. Nitshke U., Shvarts P., Krylov V.S., Linde H. Numerical method of problem solution on stationary heat- and massoexchange in two-phase system with cellular flows structure. // Theor. Osn. Khim. Tekhnol. 1985. V. 19. No. 6. P. 729–734.

122. Novick-Cohen A., Segel L.A. Nonlinear aspects of the Cahn-Hilliard equation // Physica D. 1984. V. 10. No. 2. P. 277–298.

123. Palmer H.J., Berg J.C. Hydrodynamic stability of surfactant solutions, heated from below // J. Fluid Mech. 1972. V. 51. P. 385–402.

124. Pearson J.R. On convection cells induced by surface tension // J. Fluid Mech. 1958. V. 4. No. 5. P. 489–500.

125. Pikkov L.M., Rabinovich L.M. On the calculation of masstransfere velocity in fluid in the presence of Marangoni effect //Theor. Osn. Khim. Tekhnol. 1989. V. 23. No. 2. P. 166–170.

126. Poddubnaya L.G., Rudakov R.N., Shaidurov G.F. Heat instability of a two-layer fluid in a spherical vessel // Uch. zap. Perm univ. 1968. No. 184. Hydrodynamic. Iss. I. P. 23–40.

127. Polezhaev V.I., Bune A.V., Verezub N.A. // Mathematical modelling of convective heatmassexchange on the basis of the Navie-Stoks equations. M.: Nauka. 1987. 272 P.

128. Popov V.V. Mixed convection in a two-layer fluid // Theor. Osn. Khim. Tekhnol. 1981. V. 15. No. 3. P. 398–404.

129. Potter D. Computational methods in physics // M.: Mir. 1975. 392 P.

130. Povitsky A.S., Lyubin L.Ya. Main features of dynamics and heat-massexchange of fluids and gases in microgravity // M.: Mashin-building. 1972. 252 P.

131. Projahn U., Beer H. Theoretical and experimental study of transient and steady-state natural convection heat transfer from a vertical flat plate partially immersed in water// Int. J. Heat Mass Transfer. 1985. V. 28. No. 8. P. 1487–1498.

132. Projahn U., Beer H. Thermogravitational and thermocapillary convection heat transfer in concentric and eccentric horizontal, cylindrical annuli filled with two immiscible fluids // Int. J. Heat Mass Transfer. 1987. V. 30. No. 1. P. 93–107.

133. Pshenichnikov A.F., Tokmenina G.A. Deformation of free fluid surface by thermocapillary motion // Izv. AN SSSR. Mekh. Zhidk. i Gaza. 1983. No. 3. P. 150–153.

134. Pukhnachev V.V. The appearance of thermocapillary effect in a thin fluid layer // Hydrodynamics and heat-massexchange of fluid flows with a free surface. Novosibirsk. Institute of Heat Physics SB AN SSSR. 1985. P. 119–127.

135. Pukhnachev V.V. Thermocapillary convection at reduced gravitation // Novosibirsk. Institute of Hydrodynamics SB AN SSSR. 1987. 80 P.

136. Rasenat S., Busse F.H., Rehberg I. A theoretical and experimental study of double-layer convection // J. Fluid Mech. 1989. V. 199. P. 519-540.

137. Rednikov A.E., Ryazantsev Yu.S. On the thermocapillary drop motion with heterogeneous heat release // Prikl. Math. i Mekh. 1989. V. 53. Iss. 2. P. 271–277.

138. Renardy M., Renardy Y. Bifurcating solutions at the onset of convection in the Benard problem for two fluids // Physica D. 1988. V. 32. P. 227–252.

139. Renardy Y. Interfacial stability in a two-layer Benard problem // Phys. Fluids. 1986. V. 29. No. 2. P. 356–363.

140. Renardy Y., Joseph D. Oscillatory instability in a Benard problem of two fluids // Phys. Fluids. 1985. V. 28. No. 3. P. 788–793.

141. Renardy Y., Renardy M. Pertubation analysis of steady and oscillatory onset in a Benard problem with two similar liquids // Phys. Fluids. 1985. V. 28. No. 9. P. 2699–2708.

142. Riahi N. Nonlinear Benard-Marangoni convection // J. Phys. Soc. Jap. 1987. V. 56. No. 10. P. 3515–3524.

143. Rivkind V.Ya. Stationary motion of weak deformable drop in a flow of viscous fluid // Zap. nauchn. semin. LOMI AN SSSR. 1977. V. 69. P. 157–170.

144. Rivkind V.Ya., Ryskin G.M. Flow structure at the motion of spherical drop in the liquid medium in the region of transient Reynolds numbers // Izv. AN SSSR. Mekh. Zhidk. i Gaza. 1976. No. 1. P. 8–15.

145. Rivkind V.Ya., Ryskin G.M., Fishbein G.A. Overflow of spherical drop in a transient region of Reynolds numbers // Prikl. Math. i Mekh. 1976. Iss. 4. P. 741–745.

146. Rivkind V.Ya., Sigovtsev G.S. The problem on the drop motion in heterogeneous temperature field // Hydrodynamics and heat exchange in microgravity. M.: Nauka. 1982. P. 78–82.

147. Roberts M., Swift J.W., Wagner D.H. The Hopf bifurcation on a hexagonal lattice // Multiparameter Bifurcation Theory. Providence. 1986. P. 283-318.

148. Ryazantsev Yu.S. On thermocapillary motion of reacting drop in chemically active media // Izv. AN SSSR. Mekh. Zhidk. i Gaza. 1985. No. 3. P. 180–183.

149. Savistovskii G. Interphase phenomena // Last achivements in the liquid extraction / edited by Hanson K.M.: Khimiya. 1974. P. 204–254.

150. Savistovskii G. Interphase convection // Hydrodynamic of interphase surfaces. M.: Mir. 1984. P. 194-208.

151. Scanlon J.W., Segel L.A. Finite amplitude cellular convection induced by surface tension // J. Fluid Mech. 1967. V. 30. P. 149-162.

152. Schluter A., Lorts D., Busse F. On the stability of steady finite amplitude convection // J. Fluid Mech. 1965. V. 23. No. 1. P. 129–144.

153. Schramm R., Reineke H. Numerical convection in a horizontal layer of two different fluids with internal heat sources // 6 th Int. Heat Transfer Conf. Toronto. 1978. V. 2. P. 299–304.

154. Schwabe D., Scharmann A. Measurements of the critical Marangoni number of the laminar-oscillatory transition of thermocapillary convection in floating zones // Proc. of the 5 th Eur. Symp. on Material Sciences under Microgravity. Schloss Elmau. 1984. ESA-SP-222. 1984. P. 281–289.

155. Schwabe D., Scharmann A. Messung der kritischen Marangonizahl fur den Ubergang von stationarer zu oszillatorischer thermokapillarer konvektion unter Mikrogravitation: Ergebnisse der Experimente in den ballistischen Raketen TEXUS 5 und TEXUS 8 // Z. Flugwiss. und Weltraumforsch. 1985. V. 9 No. 1. P. 21–28.

156. Scriven L.E. Dynamics of a fluid interface. Equation of motion for Newtonian surface fluids // Chem. End. Sci. 1960. V. 12. P. 98–108.

157. Scriven L.E., Sternling C.V. On cellular convection driven by surface-tension gradients: effects of mean surface tension and surface viscosity // J. Fluid Mech. 1964. V. 19. No. 3. P. 321–340.

274

158. Shliomis M.I., Yakushin V.I. Convective instability of two immiscible fluids filling spherical cavity in an arbitrary relation // Uch. zap. Perm univ. 1970. No. 216. Hydrodynamics. Iss. 2. P. 15–32.

159. Shubert G., Anderson C.A. Finite element calculations of very high Rayleigh number thermal convection // Geophys. J. Roy. Astron. Soc. 1985. V. 80. P. 575–602.

160. Shubert G., Turcotte D.L. Phase changes and mantle convection // J. Geophys. Res. 1971. V. 76. No. 5. P. 1424–1432.

161. Shvarts P., Vilke G., Krylov V.S. Analyse of hydrodynamic stability of interphase boundary in the presence of Marangoni effect // Theor. Osn. Khim. Tekhnol. 1982. V. 16. No. 6. P. 777–783.

162. Shvartsblat D.L. Steady convective motions in a flat horizontal fluid layer with permeable boundaries // Izv. AN SSSR. Mekh. Zhidk. i Gaza. 1969. No. 5. P. 84–90.

163. Simanovskii I.B. Numerical investigation of convection in a system of two immiscible fluids heated from below // Convective flows and hydrodynamic stability. Sverdlovsk. UNC AN SSSR. 1979. P. 126–130.

164. Simanovskii I.B. Finite-amplitude convection in a two-layer system // Izv. AN SSSR. Mekh. Zhidk. i Gaza. 1979. No. 5. P. 3–9.

165. Simanovskii I.B. Application of markers and cells method to the problem on convection in a two-fluid system // Convective flows. Perm. 1979. Iss. 1. P. 45–51.

166. Simanovskii I.B. On stability of two-layer systems with respect to the convective mixing // Izv. AN SSSR. 1983. No. 3. P. 118–122.

167. Simanovskii I.B. Steady and oscillatory convecive flows in a two-layer system // Teplofiz. Vys. Temp. 1985. V. 23. No. 2. P. 305–308.

168. Slavchev S.G., Kozhukharova Zh. D. Numerical modelling of thermocapillary convection in non-cylindrical liquid cristall zone // Cybernetics questions. Computational aerohydromechanics / edited by O.M. Belotserkovskii. M.: 1987. P. 113–127.

169. Smith K.A. On convective instability induced by surface-tension gradients // J. Fluid Mech. 1966. V. 24. P. 401–414.

170. Sokhranskii V.B., Cherkasheninov V.I. Underground gas-oil storages of mine type // M.: Nedra. 1978. 206 P.

171. Sorokin L.E., Tarunin E.L., Shliomis M.I. Monotonous and oscillatory regimes of media convection in a magnetic field // Magnetic Hydrodynamics. 1975. No. 4. P. 22–30.

172. Space technology / edited by Steg L. // M.: Mir. 1980. 422 P.

173. Sparrow E.M., Myrum T.A. Experiments on natural convection in complex enclosed spaces containing either two fluids or a single fluid // Int. J. Heat Mass Transfer. 1987. V. 30. No. 7. P. 1247–1258.

174. Sternling C.V., Scriven L.E. Interfacial turbulence: hydrodynamic instability and the Marangoni effect // AIChE J. 1959. V. 5 No. 4. P. 514–523.

175. Stone H.L. Iterative solution of implicit approximations of partial differential equations // SIAM J. Num. Analysis. 1968. V. 5. P. 530–558.

176. Szekely J., Todd M.R. Natural convection in a rectangular cavity: transient behavior and two phase systems in laminar flow // Int. J. Heat Mass Transfer. 1971. V. 14. No. 3. P. 467–482.

177. Takeuchi H., Sakata S. Convection in a mantle with variable viscosity // J. Geoph. Res. 1970. V. 75. No. 5. P. 921–928.

178. Tarunin E.L. Numerical investigation of free convection // Uch. zap. Perm univ. 1968. No. 184. Hydrodynamics. Iss. I. P. 135–168.

179. Technological experiments in microgravity // Sverdlovsk. UNC AN SSSR. 1983. 181 P.

180. Tomh A., Aiplt K. Numerical calculations of fields in technics and physics // M.- L.: Energiya. 1964. 208 P.

181. Thompson J.F., Thames F.C., Mastin C.W. Automatic numerical generation of body-fitted curvilinear coordinate system for field containing any number of arbitrary two-dimensional bodies // J. Comput. Phys. 1974. V. 15. P. 299–319.

182. Van Lamsweerde-Gallez D., Bisch P.M., Sanfeld A. Hydrodynamic stability of monolayers at fluid-fluid interfaces. II. Dipole interactions // J. Colloid and Interface Sci. 1979. V. 71. No. 3. P. 513–521.

183. Villers D., Platten J.K. Rayleigh-Benard instability in system presenting a minimum in surface tension // Proc. 5 th Europ. Symp. Material Sciences under Microgravity. Schloss Elmau. 1984. ESA-SP-222. 1984. P. 281–289.

184. Villers D., Platten J.K. Marangoni convection in systems presenting a minimum in surface tension // Physicochemical Hydrodynamics. 1985. V. 6. No. 3. P. 301–310.

185. Vozovoiy L.P., Nepomnyashchy A.A. Convection in a horizontal layer in the presence of space modulation of temperature on boundaries // Uch. zap. Perm univ. 1974. Hydrodynamics. Iss. 7. P. 105–117.

186. Wang R., Sanfeld A. Marangoni effect on the stabillity of two immiscible fluids in relative motion // Annales de Physique. Colloque N 2. Supplement au No. 3. 1988. V. 13. P. 105–113.

187. Welander P. Convective instability in a two-layer fluid heated uniformly from above // Tellus. 1964. V. 16. No. 3. P. 349–358.

188. Winters K.H., Plesser T., Cliffe K.A. The onset of convection in a finite container due to surface tension and buoyance // Physica D. 1988. V. 29. P. 387–401.

189. Wuest W. Thermokapillare Stromungen (Thermishe Marangonikon-vektion) // Z. Flugwiss. und Weltraumforsch. 1982. V. 6. No. 3. P. 137–146.

190. Yalamov Yu. I., Sanasaryan A.S. Drops motion in heterogeneous in temperature viscous media // Inzh. Fiz. Zh. 1975. V. 28. No. 6. P. 1061–1064.

191. Yanenko N.N. The method of fraction steps for solution of multi-dimensional problems in mathematical physics // Novosibirsk. Nauka. 1967. 195 P.

192. Young N.O., Goldstein J.S., Block M.J. The motion of bubbles in a vertical temperature gradient // J. Fluid Mech. 1959. V. 6. P. 350–356.

193. Zeren R.W., Reynolds W.C. Thermal instabilities in two-fluid horizontal layers // J. Fluid Mech. 1972. V. 53. Pt. 2. P. 305–327.

194. Zhukhovitsky E.M. On stability of ununiformly heated fluid in a spheric vessel // Prikl. Math. i Mekh. 1957. V. 21. No. 5. P. 689–693.

195. Zuev A.L., Pshenichnikov A.F. Deformation and fluid film fracture caused by thermocapillary convection // Prikl. Mekh. i Tekhn. Fiz. 1987. No. 3. P. 90–95.

INDEX